International Conference on Vibration, Sound and System Dynamics (ICVSSD 2018)

MATEC Web of Conferences Volume 217 (2018)

Penang, Malaysia
20 August 2018

Editors:

A.Z. Ahmad Mazlan
H. Endan

ISBN: 978-1-5108-7309-4

Printed from e-media with permission by:

Curran Associates, Inc.
57 Morehouse Lane
Red Hook, NY 12571

Printed by Curran Associates, Inc. (2018)

For additional information, please contact EDP Sciences – Web of Conferences at the address below.

EDP Sciences – Web of Conferences
17, Avenue du Hoggar
Parc d'Activité de Courtabœuf
BP 112
F-91944 Les Ulis Cedex A
France

Phone: +33 (0) 1 69 18 75 75
Fax: +33 (0) 1 69 28 84 91

contact-edps@webofconferences.org

Additional copies of this publication are available from:

Curran Associates, Inc.
57 Morehouse Lane
Red Hook, NY 12571 USA
Phone: 845-758-0400
Fax: 845-758-2633
Email: curran@proceedings.com
Web: www.proceedings.com

TABLE OF CONTENTS

ICVSSD 2018

2nd International Conference on Vibration, Sound and System Dynamics

20th August 2018

Olive Tree Hotel,

Penang, Malaysia.

Organized by: TheVibrationLab, School of Mechanical Engineering, Universiti Sains Malaysia (USM)

Co-organized by:

Universiti Kuala Lumpur Malaysian Spanish Institute (UniKL-MSI)

Kolej Universiti Tunku Abdul Rahman (TARC)

Universiti Tenaga Nasional (UNITEN)

SPONSORS

Platinum Sponsor

DAG Technologies Sdn. Bhd.

Silver Sponsors

Invicom Test and Measurement Sdn. Bhd.

MSC Software (S) Pte Ltd

2018 International Conference on Vibration, Sound and System Dynamics (ICVSSD 2018), TheVibrationLab, School of Mechanical Engineering, Universiti Sains Malaysia, Penang, Malaysia, 20th August 2018.

Editors: Ahmad Zhafran Ahmad Mazlan and Hishammudin Endan

INTRODUCTION

TheVibrationLab, USM is actively pursuing research in the fields of Vibration, Sound and System Dynamics Engineering. The laboratory in collaboration with UniKL-MSI, TARC and UNITEN proposes to conduct the 2nd ICVSSD conference for the year 2018 on 20th August 2018 located at the Olive Tree Hotel, Penang.

The conference is intended to bring all the researchers from universities and industries to share their knowledge in the area of research and development. It is anticipated that the conference will also be beneficial for the researchers and graduate students to improve their quality of research work. The theme of this conference for year 2018 is **" Translational Research in Knowledge Driven Economy "** to encourage presenters to highlight their research work that can lead to improved competitiveness and well-being of the industry and society.

AIM

To bring together researchers in the area of **vibration, sound, system dynamics and other related areas** to share their latest findings in the area.

OBJECTIVES

- To increase the interaction and communication between the university and industry researchers and graduate students.
- To expose the researchers and graduate students in scientific and technical communication in vibration, sound and system dynamics areas
- To promote TheVibrationLab USM research activities and encourage publication works

CHAIRMAN

Professor Dr. Zaidi Mohd Ripin

ICVSSD 2018 ORGANIZING COMMITTEE

Vice Chairman: Dr. Ooi Lu Ean

Secretary: Mrs. Norashikin Ismail

Publication: Dr. Ahmad Zhafran Ahmad Mazlan

Registration: Dr. Abdullah Aziz Saad

Technical: Mr. Wan Mohd Amri Wan Mamat Ali

Publicity: Mrs. Azni Shahida Khalid

Treasurer: Mr. Mohamad Izudin Alisah

Webmaster: Mr. Hishammudin Endan

ICVSSD 2018 TECHNICAL COMMITTEE

Prof. Dr. D.C. Barton (University of Leeds, UK)

Prof. Dr. Roslan Abdul Rahman (Universiti Teknologi Malaysia)

Prof. Dr. Musa Mailah (Universiti Teknologi Malaysia)

Prof. Datuk Ir. Dr. Mohd Jailani Mohd Nor (Universiti Teknikal Malaysia Melaka)

Prof. Ir. Dr. Jawaid Iqbal Inayat Hussain (Universiti Tenaga Nasional)

Assoc. Prof. Ir. Dr. Asan Ghani Abdul Muthalif (International Islamic University Malaysia)

Assoc. Prof. Ir. Dr. Nawal Aswan Abdul Jalil (Universiti Putra Malaysia)

Ir. Dr. Feizal Yusof (Universiti Sains Malaysia)

Dr. Chuah Hun Guan (KDU University College)

Dr. Mohamad Ezral Baharudin (Universiti Malaysia Perlis)

Dr. Ko Ying Hao (TAR University College)

Dr. Norilmi Amilia Ismail (Universiti Sains Malaysia)

Dr. Muhammad Najib Abdul Hamid (Universiti Kuala Lumpur)

Dr. Tan Wei Hong (Universiti Malaysia Perlis)

Dr. Teoh Choe Yung (TAR University College)

Dr. Zahrul Fuadi (Syiah Kuala University, Indonesia)

Dr. Ahmad Zhafran Ahmad Mazlan (Universiti Sains Malaysia)

edp sciences

Web of Conferences

Statement of Peer review

In submitting conference proceedings to *Web of Conferences*, the editors of the proceedings certify to the Publisher that

1. They adhere to its **Policy on Publishing Integrity** in order to safeguard good scientific practice in publishing.
2. All articles have been subjected to peer review administered by the proceedings editors.
3. Reviews have been conducted by expert referees, who have been requested to provide unbiased and constructive comments aimed, whenever possible, at improving the work.
4. Proceedings editors have taken all reasonable steps to ensure the quality of the materials they publish and their decision to accept or reject a paper for publication has been based only on the merits of the work and the relevance to the journal.

Title, date and place of the conference

Title: 2018 International Conference on Vibration, Sound and System Dynamics (ICVSSD 2018)
Date: 20 August 2018
Place: Olive Tree Hotel, Penang, Malaysia

Proceedings editor(s):

1) Editor in Chief: Dr. Ahmad Zhafran Ahmad Mazlan
2) Assistant Editor in Chief: Hishammudin Endan

Date and editor's signature

18 September 2018

DR. AHMAD ZHAFRAN BIN AHMAD MAZLAN
Pensyarah Universiti
Pusat Pengajian Kejuruteraan Mekanik
Kampus Kejuruteraan
Universiti Sains Malaysia

PREDICTION OF THE FLOW-INDUCED VIBRATION RESPONSE OF THE CHENDEROH DAM LEFT BANK SECTION

Mohamad Hazwan Mohd Ghazali[1], Mohd Hafiz Zawawi[2], Nurul Husna Hassan[2], Mohd Rashid Mohd Radzi[2,3], Ahmad Zhafran Ahmad Mazlan[1]*, Mohamad Aizat Abas[1] and Mohd Remy Rozainy Mohd Arif Zainol[4]

[1]School of Mechanical Engineering, Engineering Campus, Universiti Sains Malaysia, Penang
[2]Department of Civil Engineering, College of Engineering, Universiti Tenaga Nasional, Selangor
[3]TNB Generation Division, Tenaga Nasional Berhad, Kuala Lumpur
[4]School of Civil Engineering, Engineering Campus, Universiti Sains Malaysia, Penang
*Email: zhafran@usm.my

ABSTRACT

Flow-induced vibration is a common phenomenon that happened in any of dam structures during the operational condition. This includes the effect of water spilling from the upstream to the downstream of the dam due to high water volume at the upstream side. The release of water from the dam can be beneficial in generating the electricity source to the surrounding areas. However, in some cases, the spill of water can induced the significant vibration effects to the dam structure. In this study, the prediction of the flow-induced vibration response at the left bank section of the real scale Malaysian Chenderoh Dam model is simulated using the ANSYS software. The input force disturbances from the flow of the water at the left bank section during the normal water spilling condition is investigated. The results of frequency domain response and operational defection shapes (ODS) from the effect of flow-induced vibration are compared with the natural frequencies and mode shapes of the dam. From the results, the transient vibration responses due to the flow of water happened at the frequency of 13.3 Hz while the natural frequency of the left bank section occurred at 52.3 Hz, which indicates that there is no resonance phenomenon for the normal case of water spilling at the left bank section of the dam structure. This result is useful for the dam operation section in order to avoid any disaster of the dam structure.

Keywords: Flow-induced vibration, water spilling, dam structure, operational deflection shapes, resonance.

INTRODUCTION

Vibrations induced by the fluid on the dam structure can lead to the catastrophic failure if its natural frequencies coincide with the shedding frequencies of the flow. Apart from that, this vibration can also reduce the performance of structure or equipment [1]. The value of the forces induced by the fluid depends on the pressure and velocity of the flow as well as the material properties of the actual structure [2]. Studies regarding the flow-induced vibration on various samples have been conducted over the years but only few of it involved the dam structures. Research on the sliding gate dam has been conducted by Ilgar et. al., [3] to study the dynamic behavior of the gate for different shape of valves. From this study, flat valve gate is found to have the highest possibility of damage to the dam wall. Another study has been conducted by Sanchez and Salazar that aims to determine pressure distribution and velocity profile of discharge flow for radial gates opening value of 2, 5, 10 and 14 m, with 10 and 14 opening being the uncommon cases [4]. The flow rates for each opening are 12.07 m^3/s, 11.85 m^3/s, 11.85 m^3/s and 11.47 m^3/s respectively. From the simulation, there is a vortex formation for 10 m opening and also turbulence indicated at the discharge for 14 m opening. However, the results showed no sign of vibration due to the flow of water at the radial gate.

Generally, there are two types of water flow phenomenon in Chenderoh dam; water spilling and water surging. Water spilling occurs when all the gates are fully opened and maximum amount of water is released from upstream to downstream area. For the water surging, all the gates are suddenly closed during emergency cases. In this paper, the water spilling case is studied where the left bank section is modeled exactly as the original drawing of the Chenderoh dam. The FEM model used the properties of the real concrete materials and simulation study is performed using ANSYS transient structural and fluent workbenches and the results are compared with the modal and harmonic analyses.

METHODOLOGY

Modelling of the Left Bank Section of the Dam Structure

For the Chenderoh Dam, left bank section is the end part of the overall dam, where it connects with the bottom outlet section. This part is built half hollow and half solid. The three-dimensional (3D) model of the left bank section is constructed using SolidWorks software. Figure 1 shows the real and 3D model of the left bank section.

Figure 1: Picture of the real left bank section (left) and 3D model of left bank section (right)

Boundary Condition of the Fluid and the Structure of the Dam

For the left bank structure in the transient analysis, the boundary condition is where all the bottom faces are set as fixed support. The side faces that connected with the bottom outlet also been set as fixed support for this case as shown in Figure 2.

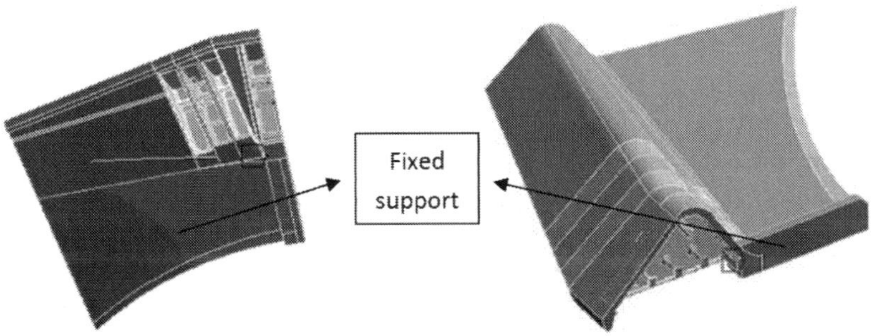

Figure 2: Boundary condition of structure at bottom (left) and side faces (right)

For the fluid, the types of boundary condition for inlet and outlet sections are velocity inlet and pressure outlet respectively. The velocity magnitude of water is set as 1 m/s. Figure 3 shows the inlet and outlet section of the left bank section.

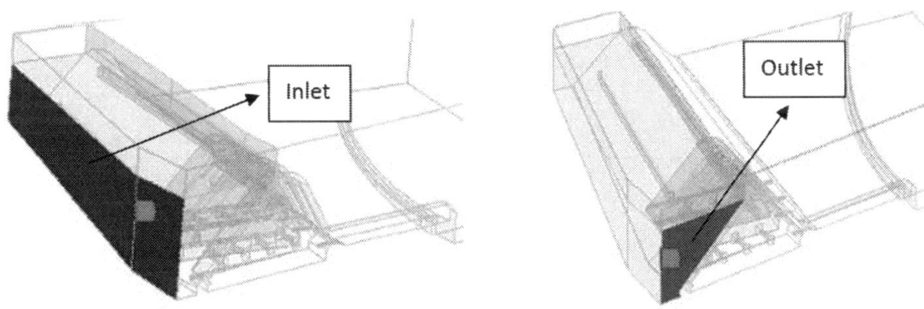

Figure 3: Inlet (left) and outlet (right) section for the left bank

Two-way Fluid-structure Interaction (FSI) of the Left Bank Section

Two-way FSI of the left bank section is made up of two systems analyses and one integrated system located in ANSYS software. The analysis systems are transient structural and fluid flow (fluent) whereas the integrated system is system coupling. All of this systems are linked together as shown in Figure 4.

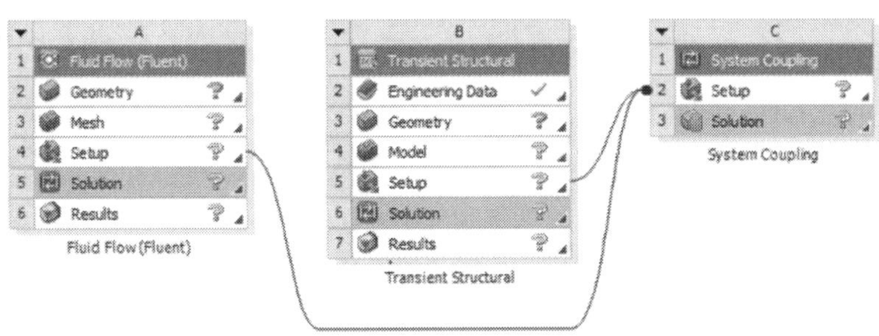

Figure 4: Linked setup between systems in two-way FSI

 https://doi.org/10.1051/matecconf/201821701001

For the transient structural system, the type of mesh used is hex dominant mesh. In the analysis of the left bank section, the step end time is set to 5 seconds and the step control is defined by substeps. The initial and minimum substep used is 1 and the maximum substep is 100. After setting-up the boundary condition as shown in Figure 2, the standard Earth gravity acceleration of 9.8 m/s^2 is applied at $-y$ direction. In the fluent system, an automatic mesh is used in order to reduce the simulation cost in terms of time and number of elements. A 2nd order upwind is chosen for the solution method of turbulent kinetic energy and dissipation rate. 0.01 second is set as a time step size with maximum iteration of 200 times. In the system coupling (integrated system), the end time is set similar to transient structural system which is 5 seconds and the time step size used is 0.01 second, similar to fluent system.

RESULTS AND DISCUSSION

ODS Results of the Left Bank Section

Figure 5 shows the ODS of the left bank section with maximum displacement of 1.18 x 10^{-3} m occurs at the upper section of the left bank. On the other hand, there is no significant effect on the apron caused by the water. It can also be observed in Figure 6, the ODS of left bank section in terms of time and frequency domain. From the frequency domain graph, the highest overall deflection occurs at an operating frequency of 13.3 Hz with maximum displacement value of 2.51 x 10^{-3} m.

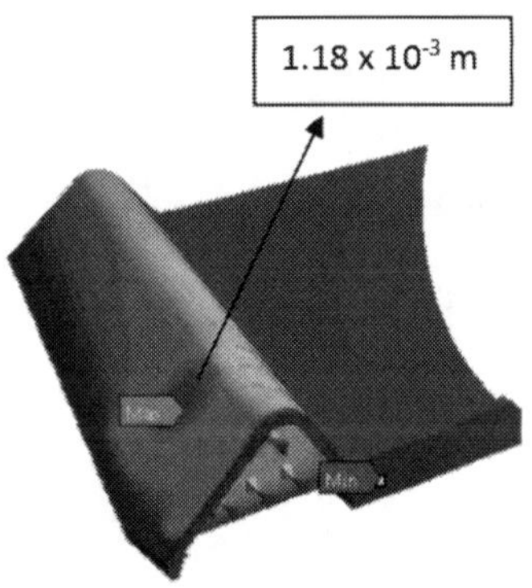

Figure 5: ODS of left bank section

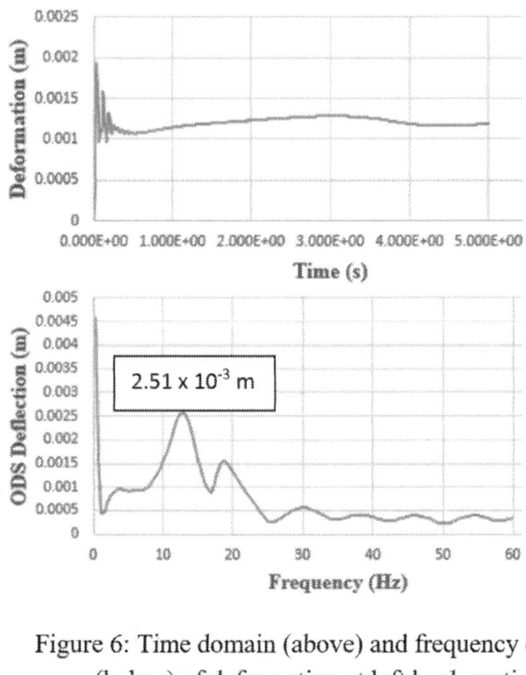

Figure 6: Time domain (above) and frequency domain (below) of deformation at left bank section

Comparison between ODS and Modal Analysis of the Left Bank Section

Figure 7 shows the comparison between ODS result from transient analysis and mode shape result from modal analysis. It can be observed that the ODS and the mode shape exhibit almost similar result where deformation occurs at the upper part of left bank section. From Figure 8, the ODS and MA deformation occurred at frequency of 13.3 Hz and 52.3 Hz, respectively, which means that they are in different frequency range. Due to the fact that these two frequencies do not coincided, there is no significant failure to the left bank section of Chenderoh Dam for the normal water spilling condition.

ODS	**Mode Shape**
Operating frequency (f_o) = 13.3 Hz Amplitude (x_o) = 2.51 x 10^{-3} m	Natural frequency (f_n) = 52.3 Hz Amplitude (x_n) = 6.96 x 10^{-4} m

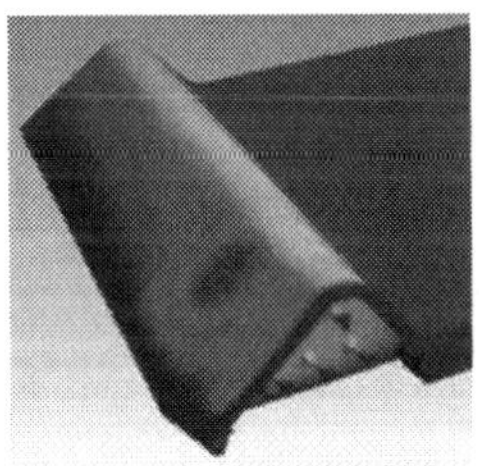

VS

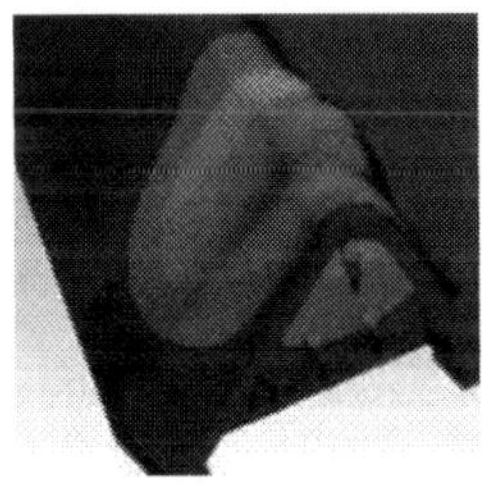

Figure 7: Transient vibration in ODS (left) and bending movement of mode shape in MA (right) of the left bank section

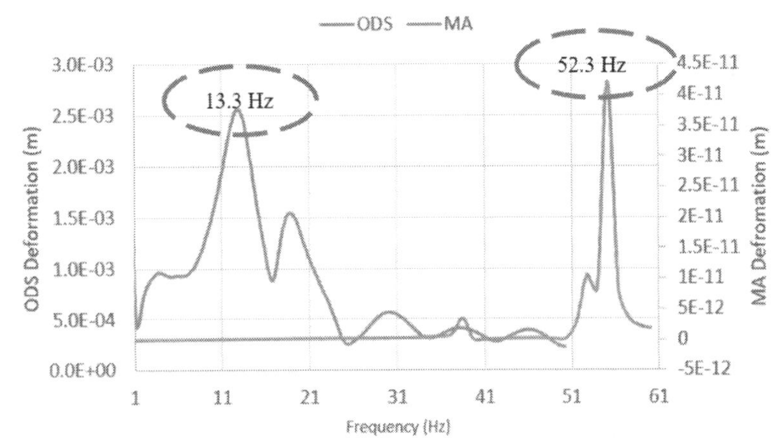

Figure 8: Comparison of MA deformation and ODS deformation in frequency domain

CONCLUSION

In this study, prediction of the flow-induced vibration response of the Chenderoh Dam left bank section has been successfully carried out. Its can be concluded that:

1) The modeling of the left bank section of the dam structure has been successfully done in SolidWorks software.

2) The two-way FSI study of the left bank section has been conducted in ANSYS software using transient structural, fluent and system coupling.

3) The ODS result of the left bank section has been achieved in terms of time and frequency domain graphs. The highest overall deflection occurs at operating frequency of 13.3 Hz with displacement value of 2.51×10^{-3} m.

4) Comparison between ODS and mode shape has been made where both deflection occurred at the upper part of left bank section and no deflection at apron. Since both deflections occurred at different frequencies, there is no critical failure to the left bank section of Chenderoh Dam.

ACKNOWLEDGEMENTS

The authors would like to acknowledge Universiti Sains Malaysia and Uniten R&D Sdn. Bhd. for providing the facilities and financial assistance under accounts (304/PMEKANIK/60313052 and U-841).

REFERENCES

[1] Virtual Prototypes (Flow-Induced Vibration). Retrieved from: https://www.ansys.com/other/virtual-prototypes/flow-induced-vibration
[2] Fluid-Structure Interaction. Retrieved from: https://www.comsol.com/multiphysics/fluid-structure-interaction
[3] Javanshir I., Javanshir N., Barmaki R. and Mahmoodi M. (2015). Modeling of the Fluid-Induced Vibrations in Sliding Gate Dams. *Journal of Vibroengineering,* Vol.17, Issue 1, p. 478-486.
[4] Sanchez D. and Salazar J. (2010). On the Effects of Water Discharge Through Radial Gates at the Caruachi Dam. *Proceedings of the ASME 2010 10th Biennial Conference on Engineering Systems and Analysis*. 10.1115/ESDA2010-24080.

FULL-FIELD VIBRATION MEASUREMENT USING CAMERA THROUGH SHEARLET SYSTEM

C. A. Wee[1*], M. Z. Ngali[1]

[1]Faculty of Mechanical and Manufacturing Engineering,
Universiti Tun Hussein Onn Malaysia,
*Email: gd160056@siswa.uthm.edu.my

ABSTRACT

Traditionally, vibration measurement is done using an accelerometer or Laser Doppler Vibrometer, which is an intrusive and single point measurement respectively. This paper demonstrated vibration signal extracted from a perceptually invisible vibrating object using only a camera non-intrusively at full-field capacity. The camera that is capable to capture 1000 frames-per-second video was used to capture the motion of the vibrating object. Each frame of the video was decomposed using complex shearlet transform and the extracted signal was compared to an accelerometer. Shearlet decomposed each frame of the video into complex coefficients which were later used to recover the motion between two consecutive frames. Phase information that were used to retrieve the vibration signal were weighted to decrease unwanted noise. Resonant frequencies of a simply supported beam at 104.0 Hz, 209.5 Hz and 396.0Hz were successfully recovered. Single frequency extraction from an excited loud-speaker at specific frequencies were also conducted and showed a clear peak-to-valley frequency spectrum recovery. The potential of using camera as a full-field displacement measurement where each pixel acting as a vibrometer was explored. Full-field test to recover the mode shape of a circular membrane showed promising result with eight mode shapes successfully retrieved. The experiments proved that each individual pixel was able to retrieve motion at subpixel level that is at 0.00001 pixel scale.

Keywords: full-field, vibration measurement, camera, mode shape, shearlet, subpixel motion.

INTRODUCTION

Overview

Vibration of an object is traditionally measured using accelerometer that requires the sensor attached to the vibrating body. While accelerometer is accurate and precise in measuring displacement, it is intrusive to the measuring object or structure itself as the mass of the accelerometer somehow altered the dynamic response. There are devices that uses non-intrusive method to overcome this problem. Laser Doppler Vibrometer (LDV) projected a laser beam onto the surface of a vibrating surface and measure its velocity response from Doppler shift effect[1]. It has high spatial resolution and could be easily set-up. However, LDV is a single point axial vibrometer as it only measure one point at a time. Continuous scanning laser Doppler vibrometer (CSLDV) have been introduced to overcome the limitation of LDV. CSLDV employed the LDV whose laser beam is sweeping across the surface of vibrating object to measure the vibration at multiple points.

Recently, advanced full-field measurement methods had been introduced [2-4] by using high speed camera. These techniques involved using a high frame rate camera to capture a series of images at thousandth of second per frame. Each frame is then decomposed into Fourier domain using complex steerable pyramid (CSP) [5] at multi-resolution and multi-orientations. The phase of the decomposed image was then processed and filtered to reduced noise and extracted the encoded vibration signal in its phase. This method is highly efficient and accurate in extracting sub-pixel motion as compared to optical-flow method such as Lukas-Kanade[6] and Horn-Schunck[7] which tends to be affected by the noise as estimation operations were performed spatially. In contrast, CSP operations are all in Fourier domain.

Despite quite a number of studies had been done on the full-field measurement method, it still has many potential to be improved and explored. This paper is thus written to address the lacking. A more sophisticated method in decomposing the image using Shearlet transformation is introduced to improve the vibration signal extraction from a subpixel motion video.

This text is organized as follows: Section 2 we discussed the basic of phase-based motion estimation. Section 3 explained wavelet and shearlet system. Section 4 discussed the experiments set-up. Section 5 showed experiment result in validating the proposed method and obtaining the modal shape of a circular membrane.

Phase-Based Motion Estimation and Shearlet System

Phase-based motion estimation is a method that employ image processing technique to decompose an image into phase and magnitude domain using wavelet basis functions. Wavelet bases are well localized in space as opposed to Fourier bases that sparse in representation[8] and thus it is suitable to be used in full-field displacement measurement. It is not to be confused with phase correction [9] which estimates the motion of two images globally by computing the Phase-Only-Correlation (POC) and finding the peak of shifting using cross-power spectrum.

Well established wavelet image decomposition is steerable pyramid[5]. In [10], the authors used CSP to magnify motion in a video that is invisible to naked human eyes, while *Davis* et al. [3] successfully extended the work to recover sounds from high speed video. Steerable pyramid uses steerable Gabor-like wavelets as the basis function to represent image coefficients in multi-resolution and multi-orientations. In complex steerable pyramid, the analytic signal is made up from complex-valued sinusoids [11]. Real and imaginary parts of the coefficients have more advantage over real-valued steerable pyramid. For instances, the magnitude of a complex coefficients are shift-invariance and its phase contain more image features especially edge. The phase in complex coefficients near an edge varies linearly with its distance to the edge[12].

Mathematically, CSP decomposed image I into different complex-valued spatial sub-bands $S_{r,\theta}$ corresponding to scale r and orientation θ. The basis functions during this transformation are scaled and oriented accordingly. At each sub-band, a complex-valued image is defined in terms of amplitude A and phase φ, by abusing of notation

$$A(r,\theta,x,y)e^{i\varphi(r,\theta,x,y)} \tag{1}$$

To extract the vibration signal between two consecutive images, local phase different between them is calculated. The local phase different φ_d between an image at time t and a reference image t_0 is defined as

$$\varphi_d(r,\theta,x,y,t) = \varphi(r,\theta,x,y,t) - \varphi(r,\theta,x,y,t_0) \tag{2}$$

However such operations will involve the phase different of unwanted noise too. Fortunately, the amplitude from CSP provide us the details about the noise. High amplitude corresponds to low noise and vice-versa. Making use of such information, we were able to filter and weight the noise. It is worth to mentions that such operation is valid for small motion [13], which this paper is focused on. Frames were then decomposed using shearlet transform. It was proved that it achieved better result than the conventional wavelet transform [14]. Wavelet tends to treat different axes equally, such property resulted in poor anisotropic feature detection [15]. Since image is a 2D signals, such behavior will have impact on the image decomposition of highly texturized image with a lot of edges. This is where shearlet system outperform the wavelet as it is highly scalable and shear-able. However classic shearlet suffer from ineffectiveness of computation when the shear parameter is large. In modern cone-adapted shearlet system [15], frequency domain is separated into two region: low-pass residual at center and two conic regions to overcome the problem of large shearing parameters as shown on the right of Figure 1.

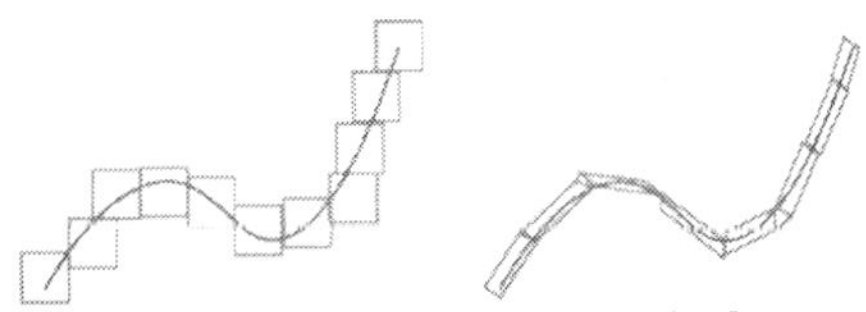

Figure 1. (left) Isotropic and; (right) Anisotropic feature approximation.

Large shearing occur when the line is near to horizontal, without cone-adapted method, multiple times of transformation are needed to fit the line perfectly. With the modern shearlet system, complex coefficients of an image in level r, cone k and orientation θ is therefore defined as

$$A(r,k,\theta,x,y,t)e^{i\varphi(r,k,\theta,x,y,t)} \tag{3}$$

and its respective phase different is

$$\varphi_d(r,,k\,\theta,x,y,t) = \varphi(r,k,\theta,x,y,t) - \varphi(r,k,\theta,x,y,t_0) \tag{4}$$

Figure 2 showed that shearlet system is well localized compared to steerable pyramid. Image discontinuities such as the edge of the umbrella could be seen clearly in the shearlet decomposed phase image.

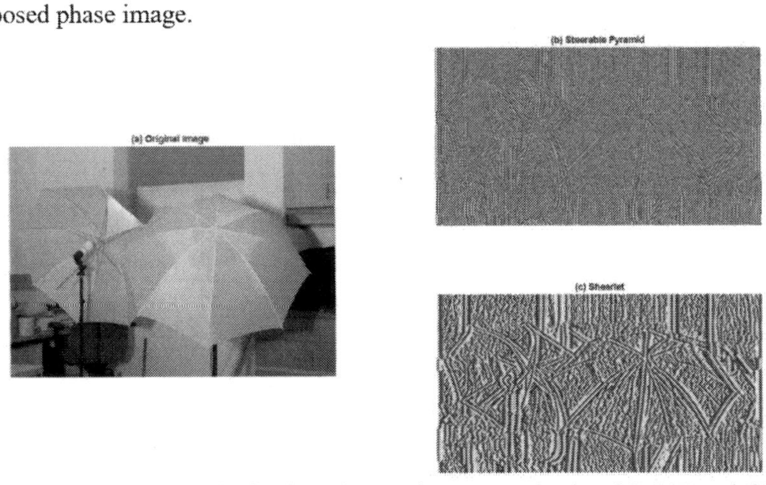

Figure 2. (a) Umbrella image [16]. Phase image decomposed using (a) CSP and (b) shearlet.

EXPERIMENTS

Employing equation (4) to calculate motion signal, we were able to extract the vibration signal for a freely hanged steel bar as shown in Figure 3. However, equation (4) is not noise resistant. Fortunately, the amplitude $A(r, k, \theta, x, y, t)$ from shearlet coefficients provided us with the details of noise. Low image contrast usually have high noise of phase. Amplitude of the image coefficients is a simple measurement of image contrast, where high amplitude indicates low image contrast and vice-versa. Thus, high amplitude corresponds to low noise. With such information, we were able to filter and weight the noise in the phase.

The camera was pointed perpendicular to the steel bar to capture all vertical motion produced by the steel bar. Two seconds of video at 1000 frame-per-second at an effective pixel of 1136x384 were recorded. A total of 2000 frames were generated from the video and stored in Bitmap (BMP) format to conserve its original image details. Approximately 12 gigabytes of data were generated from a single video alone. The frequency limit that we were able to extract from this video is 500Hz, in accordance to the Nyquist Theorem.

In order to obtain a relative motion between two frames, one could use a reference frame from a still image or by averaging the image intensities of a video over time. In our paper, we have chosen the later. Since it is not always possible to obtain a still reference image of a vibrating object in real life, thus it would be more practical to obtain a reference frame of averaged intensities over time. Besides, by averaging the frames over time, the final image would be less disturbed by noise. Such method is possible as the motion considered in this paper is at subpixel level and averaging the intensities will not change the average intensity of a pixel.

As shearlet system decomposed the frame into multi-resolution and multi-orientation, we decided to use the finest level of the coefficients as it contains more high frequency information. As for the orientation, we have chosen the direction that is parallel to the motion of the vibrating object.

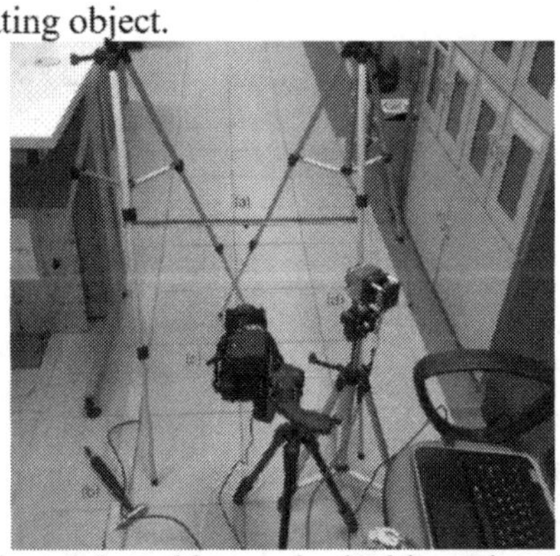

Figure 3. (a) Freely hanged steel bar on both sides using thin ropes. (b) Impact hammer used to subtly excite the steel bar. (c) Sony RX10M3 camera. (d) LED to light up the recording scene.

Next, we tested our method in a signal frequency scenario. A loud speaker is excited to produce a frequency at 225Hz, 300Hz, 425Hz and 475Hz. The amplitude is set to as lowest as possible so as to produce a motion that is subtle and invisible to human eyes. The camera was pointed fronto-parallel to the speaker so that the motion of the diaphragm is moving out of the image to the right.

Finally, we proceed with the full-field testing of camera as a multi-sensor vibrometer. A piece of latex membrane with stripes was mounted onto a PVC tube of diameter 15cm, as shown in Figure 4. The membrane is then placed in front of a loud speaker to be excited.

Figure 4. (left) A frame from the captured video and; (right) piece of latex membrane mounted on a PVC tube.

RESULTS AND DISCUSSION

The peaks of frequency spectrum obtained from our method matched closely with the value from accelerometer, as shown in Figure 5. Three resonant frequencies of the steel bar had been recovered successfully at 104.0Hz, 209.5Hz and 396.0Hz. However, resonant frequency at 347.7Hz and 356.0Hz were buried in noise in our measurement.

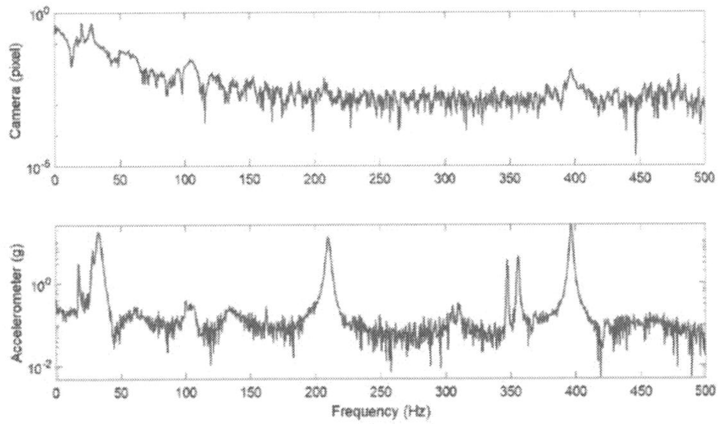

Figure 5. Frequency spectrum for camera-based and accelerometer measurement.

For the peaks exist below 50Hz, we verified that it was the unavoidable vibration that were caused by uncontrollable surrounding, such as the ventilation system.

In a single frequency test, our method was able to extract the frequency in clear peak as shown in Figure 6. This showed that our method have high signal-to-noise ratio (SNR).

225Hz

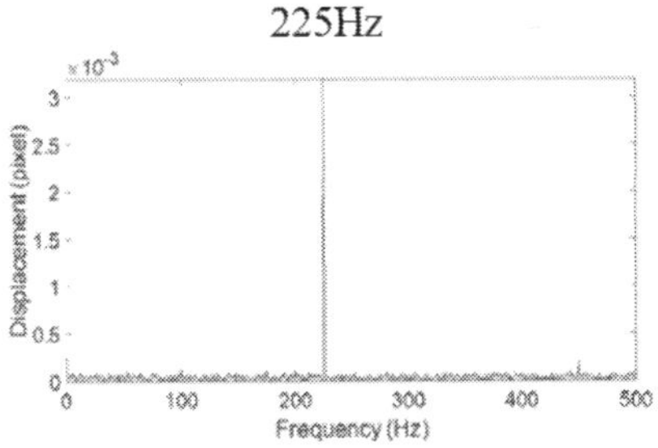

300Hz

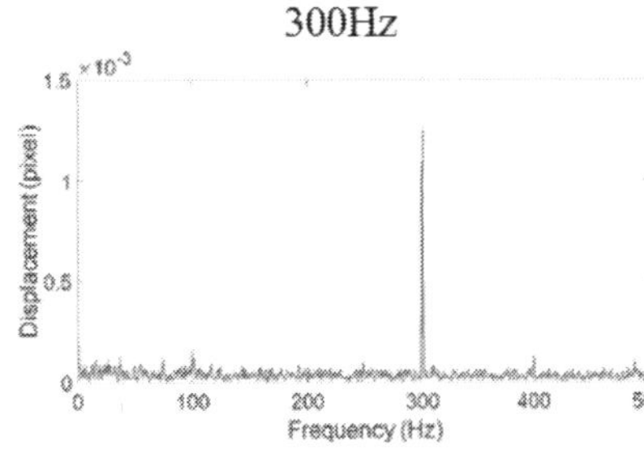

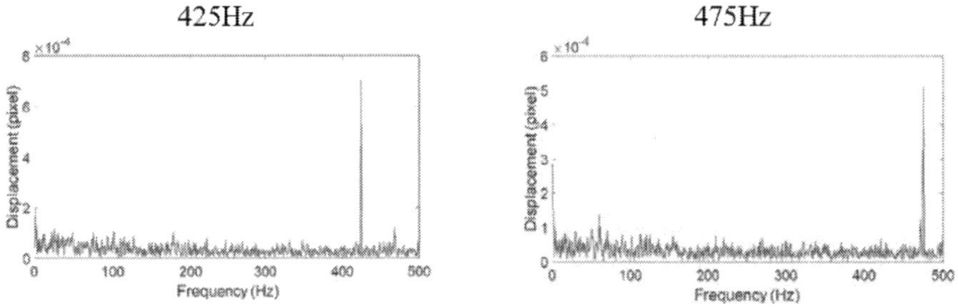

Figure 6. Frequency spectrum for single frequency test.

For our full-field test, we were able to recover eight different mode shapes at different frequencies as shown in Figure 7. The displacement of the membrane is lower than 0.00003 pixel. The conversion factor for pixel displacement to millimeter we used was 0.12 mm/pixel.

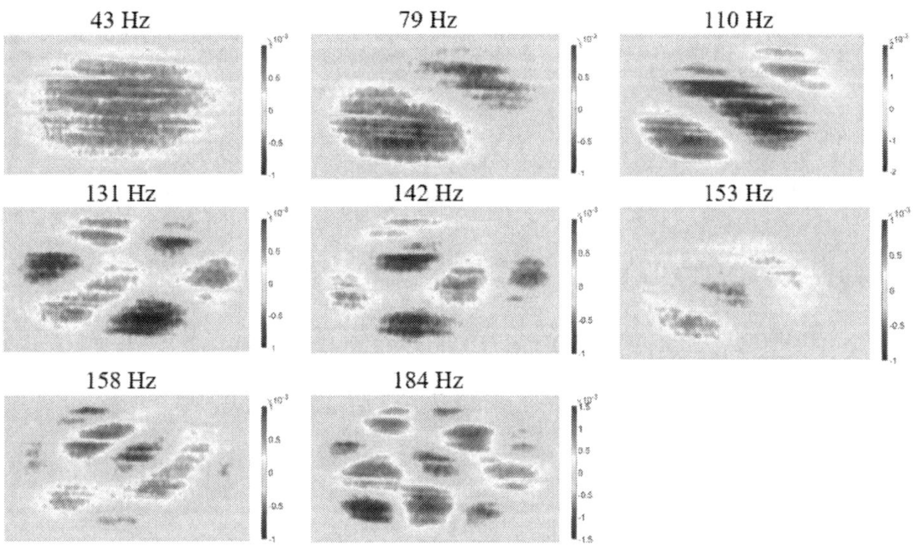

Figure 7. Recovered mode shapes at pixel displacement scale.

CONCLUSION

This research demonstrated that camera could be used as a full-field vibrometer to measure tiny vibration that is perceptually invisible to human eyes. Every pixel in the camera sensor acted as an individual vibrometer sensor. By using shearlet image decomposition, our algorithm was able to extract sub-pixel level motion that is at 0.00001 pixel scale. We successfully showed that the mode shape of a membrane at multiple frequencies were retrieved using the method we proposed and proved our statement that a camera could be used as full-field vibrometer.

ACKNOWLEDGEMENT

This research is supported by the Fundamental Research Grant Scheme (FRGS), Grant No: FRGS/2/2015 Vot 1543 by Ministry of Higher Education, Malaysia.

REFERENCES

[1] P. Castellini, M. Martarelli, and E. Tomasini, "Laser Doppler Vibrometry: Development of advanced solutions answering to technology's needs," *Mechanical Systems and Signal Processing,* vol. 20, no. 6, pp. 1265-1285, 2006.

[2] A. Davis, K. L. Bouman, J. G. Chen *et al.*, "Visual vibrometry: Estimating material properties from small motions in video." pp. 5335-5343.

[3] A. Davis, M. Rubinstein, N. Wadhwa *et al.*, "The visual microphone: passive recovery of sound from video," 2014.

[4] D. A. Ehrhardt, M. S. Allen, S. Yang *et al.*, "Full-field linear and nonlinear measurements using continuous-scan laser doppler vibrometry and high speed three-dimensional digital image correlation," *Mechanical Systems and Signal Processing,* vol. 86, pp. 82-97, 2017.

[5] E. P. Simoncelli, and W. T. Freeman, "The steerable pyramid: A flexible architecture for multi-scale derivative computation." pp. 444-447.

[6] B. D. Lucas, and T. Kanade, "An iterative image registration technique with an application to stereo vision," 1981.

[7] B. K. Horn, and B. G. Schunck, "Determining optical flow," *Artificial intelligence,* vol. 17, no. 1-3, pp. 185-203, 1981.

[8] S. Mallat, *A wavelet tour of signal processing*: Academic press, 1999.

[9] Y. Yu, and J. Wang, "Highly Accurate Estimation of Sub-pixel Motion Using Phase Correlation." pp. 186-193.

[10] N. Wadhwa, M. Rubinstein, F. Durand *et al.*, "Phase-based video motion processing," *ACM Transactions on Graphics (TOG),* vol. 32, no. 4, pp. 80, 2013.

[11] I. W. Selesnick, R. G. Baraniuk, and N. C. Kingsbury, "The dual-tree complex wavelet transform," *IEEE signal processing magazine,* vol. 22, no. 6, pp. 123-151, 2005.

[12] R. Anderson, N. Kingsbury, and J. Fauqueur, "Determining multiscale image feature angles from complex wavelet phases." pp. 490-498.

[13] T. Gautama, and M. Van Hulle, "A phase-based approach to the estimation of the optical flow field using spatial filtering," *IEEE Transactions on Neural Networks,* vol. 13, no. 5, pp. 1127-1136, 2002.

[14] H. K. Sevindir, and C. Yazıcı, "Comparison of Wavelet and Shearlet Transforms for Medical Images," *Appl. Math,* vol. 10, no. 4, pp. 1447-1452, 2016.

[15] G. Kutyniok, and D. Labate, "Introduction to shearlets," *Shearlets*, pp. 1-38: Springer, 2012.

[16] D. Scharstein, H. Hirschmüller, Y. Kitajima *et al.*, "High-resolution stereo datasets with subpixel-accurate ground truth." pp. 31-42.

MODAL ANALYSIS OF VERTICAL WIND TURBINE BLADE

Lee Zhou Yi, Choe-Yung Teoh*

Faculty of Engineering,
Tunku Abdul Rahman University College,
*Email: joeyleezy95@gmail.com
Phone: +6012-5258373

ABSTRACT

Wind turbines cannot simply be installed in Malaysia due to low wind speed condition. The project has analyzed the existing wind turbine blade (Aeolos-V 1k) design based on modal properties using computational approach (ANSYS Workbench) and redesign it. The modal analysis is simulated to observe natural frequency and corresponding mode shaped of the system under free vibration. The flow induced vibration can cause blade failure due to resonance or fatigue. Fluid Structural Interaction (FSI) ANSYS is used to the determined the interaction between the wind flow and the blade. Harmonic Response ANSYS is used to analyze the frequency response of the blade under wind induced vibration. After modification, the first mode has increased from 91.42 Hz to 102.12, since it is more than 50.92 Hz (Turbine maximum operating frequency), resonance would not occur during operating condition. The Aeolos-V's blade has been modified by using teak wood material and redesign the blade for weight reduction and aim for lower blade cost. The weight of modified blade has reduced 72.8 % after using teak wood and the efficiency of the wind turbine also increased. Modified design has been tested under Malaysia maximum wind speed of 9.44 m/s, the yield stress of teak wood (10.3 MPa) is higher than the maximum stress (4.2 MPa) obtained under force vibration which gives safety factor of 2.4. Hence, modified blade is reliable, efficient and more economic for Malaysia.

Keywords: Vertical Wind Turbine Blade, Flow Induced Vibration, Wind Energy, ANSYS Simulation.

INTRODUCTION

Renewable energy has lower impact on the environment because they are clean energy. However, there are no statistical data shown that Malaysia has installed any wind farm. The main reason of this issue is due to low average wind speed in Malaysia. Mersing has the maximum wind speed of 9.44 m/s in Malaysia [1]. To ensure that wind turbine blade designed can withstand the maximum wind speed in Malaysia. Hence, the operating frequency for designed wind turbine blade is going to be analyzed with 9.44 m/s wind load. There are two types of wind turbine which are horizontal axis wind turbine (HAWT) and vertical axis wind turbine(VAWT). VAWT are well design to harness energy in turbulent flow and low wind speed, which is more applicable for Malaysia wind condition. Darrieus rotor with straight blades called H-rotor. H- rotor has blades with airfoil cross-section as seen from the top view of turbine. H-rotor wind turbine can be design for higher tip speed ratio [2]. Force vibration occur if a system is continuously induced by an external force. Wind induced is the external force that allowed the wind turbine blade rotates to generate electricity from mechanical movement. Resonance occurs. when the operating. frequency approaches to the natural. frequency of a system. The result will produce attendant growth of vibration amplitude. Vibration resonance can cause serious damage to the object or it shorten an object's life usage.

METHODOLOGY

Aeolos - V 1k is a vertical axis wind turbine blade selected from Aeolos Wind Turbine Company for the project testing modal. The mechanical specification of testing modal is listed in Table 1. The modal is selected because its rated wind speed of 10m/s is the closest wind speed required to Mersing average wind speed, 5.4m/s. The selected design is the least bulky structure in the wind turbine market, as higher rated power required higher rated wind speed. Modal analysis is used to identify the vibration characteristic of a system when the system is under operating condition. Five modes shapes and natural frequencies are selected to simulate modal properties of Aeolos-V's blade in ANSYS Modal.

Table 1. Specification of Aeolos-V-1k

Aeolos-V 1k	Specification
Type of Wind Turbine	Giromill Darrieus
Number of blades	3 pieces
Blade material	Aluminium Alloy
Blade profile	NACA0018
Rotor radius	1 m
Span length blade	2.8 m
Chord length of the blade	0.37 m
Pitch angle	8
Height of the wind turbine	12 m
Rated wind speed	10 m/s
Tip speed ratio	3.36
Testing max. wind speed	24 m/s (Italy)

Wind energy is extracted from turbulent flow, the energy conversion can be horizontal axis wind turbine or vertical axis wind turbine. Model used in Fleunt Setup is K-epsilon Realizable with transient time solver. In this project is focusing on VAWT. The wind flow toward wind turbine blade is turbulence and while spinning the blades it creates more turbulence K-epsilon. Realizable model is used for turbulence flow specially on strong pressure flow, rotating and recirculating flow. The left side of the boundary wall is velocity inlet with 24 m/s velocity magnitude. The rest of the boundary wall are pressure-outlets, the boundary is initialized and the wind load is calculated. Fluid

Structure Interaction Analysis used to analyze the effect of aerodynamic forces exerted on a solid structure with maximum wind force at 24 m/s. The wind flow is simulated the with specified wind speed in ANSYS Fluent. Wind load is then imported in Static Structural ANSYS to simulate the stress field wind load interaction with wind turbine blade structure. Harmonic response analysis used to determine the effect of loading in term of steady-state response of a linear structure. Force vibration is obtained from Static Structural ANSYS, the maximum equivalent stress generated in

Static Structural ANSYS is consider as applied pressure on the wind turbine blade. The natural frequencies also obtain from Modal ANSYS.

The Aeolos-V's wind turbine blade is modified with different materials shown in Table 2. The purpose of modifying Aeolos-V's wind turbine blade is to reduce wind turbine blade weight, since Malaysia wind speed is low. Wind turbine blade does not need to withstand strong wind force and certain material of the blade can be reduced.

Table 2. Mechanical properties of blade materials [3].

Material	Aluminium Alloy	Carbon Fibre	Teak Wood
Density(kg/m^3)	2770	1600	581
Young's Modulus (MPa)	71000	70000	420
Poisson Ratio	0.33	0.10	0.02
Blade Weight (kg)	127.49	73.64	34.98

Three suggested hollow designs of wind blade are tested in Modal ANSYS shown in Figure 1. The blade profile still using NACA 0018 to remain the aerodynamic characteristic of Aeolos's wind turbine blade.

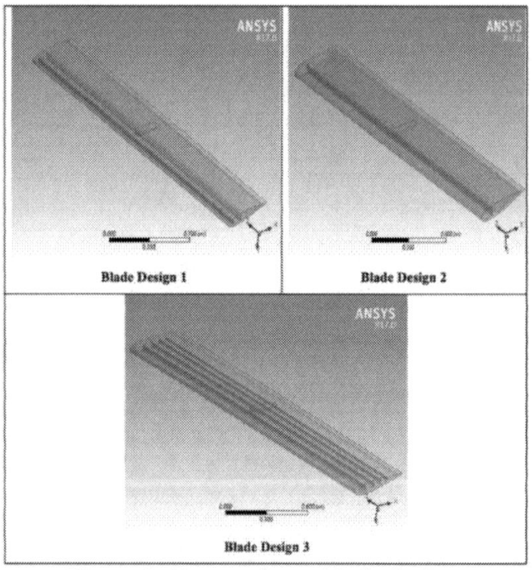

Figure 1. Types of new blade design.

RESULTS AND DISCUSSION

Modal analysis shows different natural frequency has different mode shapes in Figure 2 with reference of previous researcher [4]. First mode and second mode have flap-wise vibration characteristic. Second mode is the most. critical mode because the bending of each end of the blade are in same direction. Other modes are edge-wise mode and complex mode where they are not as critical as the first two modes. This is because does not consume much energy during wind turbine operation.

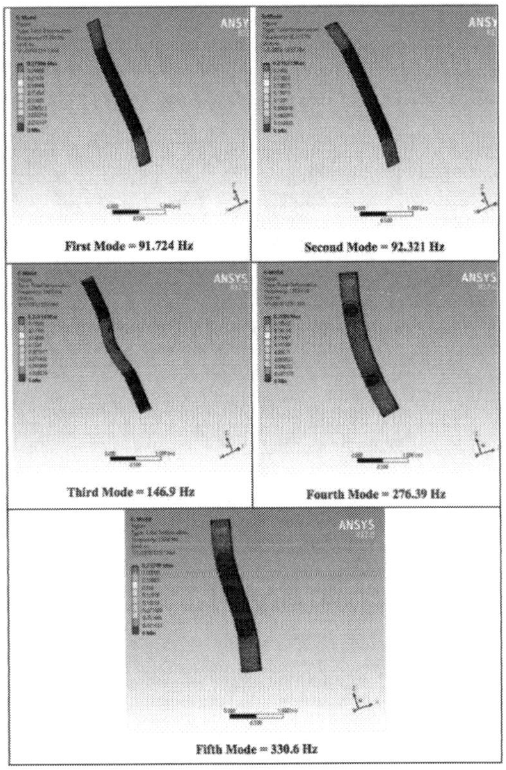

Figure 2. Mode shape for first five natural frequency of Aeolos-V's blade.

When the wind turbine blade is rotating, wind force exerted perpendicularly on the surface of the blade. The bending mode in flap-wise direction absorb more wind force more compared the other modes. The equivalent stress of 2.35 MPa is used as the pressure field on the blade that cause vibration. The result of stress is insert in the harmonic response analysis to determine the operation frequency of the blade. The effect of stress and deformation under force vibration is generated according to each frequency response in Table 3. Table 3 also shows the results of each operating frequency under wind speed of 24 m/s.

Table 3. Mode shapes, equivalent stress, deformation of the frequency response.

Modes	Mode Shape	Frequency (Hz)	Equivalent Stress (MPa)	Deformation (mm)
1	Flap-wise	91.72	12.80	0.9444
2	Flap-wise	92.32	15.21	1.118
3	Complex	146.90	16.20	0.6931
4	Edge-wise	276.39	34.70	0.3139
5	Complex	330.60	4.50	0.0027

Second mode is the most critical mode as the blade deform inward in Figure 3. The end of the blade tip has the maximum deformation. The maximum stress is at the supports of the blade. Flap-wise vibration is more critical as less amount of stress can create high deformation. The fifth mode has the least mode of 4.5 MPa and it also can the least deformation in Figure 3. Different wind load excited different frequency response which can affect the stress of the blade. Under wind load of 24 m/s results high stress in fourth mode and least stress in fifth mode.

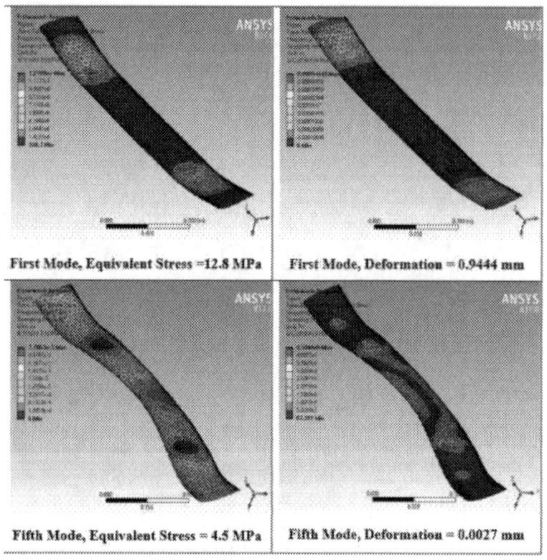

Figure 3. Equivalent stress and deformation of first mode and fifth mode.

In common wind turbine market, most of the company such as Aeolos used aluminum alloy because it can withstand higher wind speed and is less expensive. Malaysia has maximum wind speed of 9.44m/s in Mersing. The modal properties of new types of material such as carbon fibre and teak wood is listed in Table 4. The natural frequency of carbon fibre blade and teak wood blade first mode more than 50.92Hz. It means that after change to new material, the aerodynamic characteristic can be remained as Aeolos-V's blade. Teak woods are have lighter weight of 34.98kg and cost reduction. Hence, teak wood is selected as new material. The first mode of the new blades with one support in table are lower than operating frequency of wind turbine blade (50.93Hz) in Table 4. Hence, it is too risky to remove one support. Design 3 with two supports has significantly higher first mode as compare to blade with one support. Blade Design 3 are more manufacture friendly.

Table 4. modal results for different materials and blade designs.

Modes	Aluminium Alloy	Carbon Fibre	**Teak Wood**	Blade Design 1 with one support	Blade Design 2 with one support	**Blade Design 3 with one support**
1	91.72	116.81	102.12	26.983	28.362	31.994
2	92.32	117.42	102.72	42.53	43.174	38.045
3	146.90	188.34	164.1	100.77	101.57	96.475
4	276.39	375.76	317.13	150.13	145.2	164.11
5	330.60	443.62	377.46	183.69	180.49	204.55

After operating frequency response analysis tested under maximum wind speed of 9.44m/s, the maximum stress obtains from wind induced vibration is 4.2 MPa. the yield stress for teak wood is 10.3 MPa. The safety factor of the modified blade is 2.4. The design is safe to used under Malaysia maximum wind speed. The weight of the blade has been reduced as compared to Aeolos-V's blade. The weight reduced 72.8% from the Aeolos-V' Blade. The main reason is because low density of teak wood. As the weight of the blade reduce, the efficiency of the wind turbine increases. Hence, the modified design of vertical wind turbine can be applied at Mersing, Malaysia

CONCLUSION

Flow induced vibration is vital concern when design wind turbine blade. Modal properties of Aeolos-V blades have determined. The natural frequency and the corresponding mode shapes has been determined in Modal ANSYS. The first mode is the most critical mode as it has flap-wise bending mode. Other mode such as edge-wise mode and complex mode are less critical because these modes absorb less wind energy during operating wind turbine blade. The modified blade after undergoing modal analysis and operating frequency analysis, the results have shown that it would not have resonance effect and fatigue problem when applied in Malaysia. The modified blade are environmental friendly, low cost and light weight. It is very suitable to be use under low wind speed condition. Hence, Malaysia has potential to installed wind turbine as new form of renewable energy source with the modified blade. The weight has been reduced 72% from 127.49kg to 34.96kg. As the weight of the blade reduce, the efficiency of wind turbine increases. The maximum stress obtained from the blade under wind induced vibration is 4.2 MPa. The yield stress of teak wood is 10.3 MPa and factor of safety is 2.4. Hence, the modified blade can be applied in Malaysia without having fatigue failure.

ACKNOWLEDGEMENTS

I would like to acknowledge Tunku Abdul Rahman University College for providing a good environment and facilities for me to perform my studies and research. I would also like to thank my supervisor Dr. Teoh Choe-Yung who has been giving me support and guide me throughout this project.

REFERENCES

[1] Islam, M.R., Saidur, R. and Rahim, N.A., 2011. Assessment of wind energy potentiality at Kudat and Labuan, Malaysia using Weibull distribution function. Energy, 36(2), pp.985-992.

[2] Bhutta, M.M.A., Hayat, N., Farooq, A.U., Ali, Z., Jamil, S.R. and Hussain, Z., 2012. Vertical axis wind turbine–A review of various configurations and design techniques. Renewable and Sustainable Energy Reviews, 16(4), pp.1926-1939.

[3] Babu, K.S., Raju, N.S., Reddy, M.S. and Rao, D.N., 2006, June. The material selection for typical wind turbine blades using a MADM approach & analysis of blades. In Proceedings of 18th International Conference on Multiple Criteria Decision Making (MCDM 2006), Chania, Greece, June (pp. 19-23).

[4] Lemu, H.G., 2015. Advances in numerical computation based mechanical system design and simulation. Advances in Manufacturing, 3(2), pp.130-138.

THE EFFECT OF LUBRICATION ON THE VIBRATION OF ROLLER BEARINGS

N.F.M. Yusof* and Z.M. Ripin

School of Mechanical Engineering
Universiti Sains Malaysia
*Email: mefarhana@usm.my

ABSTRACT

Proper lubrication is crucial to ensure smooth operation in machineries. In rolling bearing, the improper lubrication may induce high friction and vibration level due to metal to metal contact between the rolling elements. In this study, the roller bearings with and without lubrication are investigated. The natural surface degradation of the roller bearing is monitored and the surface roughness is measured for the lubricant film thickness calculation. The film thickness is determined by the Hamrock-Dowson equation which showed that the grease lubricated bearing operated under the elastro-hydrodynamic lubrication, with the ratio of lubrication film thickness to the surface roughness of λ in the range of 0.9 to 3.65. The un-lubricated bearing was damaged after 20 minutes whereas the grease lubricated bearing continued to operate for 6600 minutes. The observation under microscope showed that the surface underwent smoothening process where the surface roughness decreases initially (running-in state) followed by roughening at the steady state where the surface roughness increases. At damage, the value of $\lambda = 0.9$ can be associated with the high level of the bearing vibration. The increase of vibration level becomes rapid at the critical value of $\lambda = 1.6$. As such the overall vibration level of the bearing can be related to the surface degradation and low film thickness.

Keywords: Roller bearing, vibration, surface roughness

INTRODUCTION

Rolling element bearings have widespread applications particularly in industrial machineries. These bearings are considered as critical mechanical components in industrial applications and the failure to detect the presence of defects may lead to catastrophic failure of the machinery. The condition monitoring of the rolling bearing is commonly carried out by vibration measurement. The rolling element bearing produces noise and vibration because of the sliding or rolling contact of the bearing components [1]. Additional factors influencing the noise and vibration include manufacturing inaccuracies, improper installation, operating speed, supported load and lubricant [2]. The trend of vibration acceleration shows sudden increase and decrease during the early cycle (running-in) of rolling contact [3]. The fluctuation of the signal level was attributed to phenomenon known as healing which is described as smoothing of sharp edge of surface asperities by continuing rolling contact [3, 4].

Lubrication plays an important role in order to ensure a smooth bearing operating condition. The function of lubrication is to form a layer of a different material between the surfaces that reduce the friction force between them. FAG Technical

Publication [5] stated that 80 % of bearing failures are caused by lubricant failure. The changes in lubrication condition of rolling bearing mainly affect its vibration in high frequency band [6, 7]. The vibration level increases with the decrement of the minimum film thickness and lambda value [8]. The higher vibration level was excited with the increase of grease base viscosity [9]. Improper lubrication can be expressed by the increase in the spectral components at bearing cage and ball spin frequency [10]. The full film lubrication (hydrodynamic lubrication) cannot be reached under certain operating conditions such as low speed, very high load, and at very high temperature as a thin and inadequately separating lubricating film is formed [11]. The boundary between fully flooded and starved regime is determined by volume, contact dimensions, oil viscosity and speed [12]. In grease-lubricated contacts, a fully flooded condition occur initially then shift to starved condition in a short time [13, 14]. The mechanical aging process reduces the thickener material subsequently decrease the grease film thickness at very low speeds, however no significant difference was found at higher speeds where the grease film thickness still approached the calculated base oil film thickness [15]. In this study, vibration level and lubrication condition of the grease lubricated bearing will be monitored until failure. The lubricant film thickness is calculated by using surface roughness value.

METHODOLOGY

Experimental Setup

A detachable roller bearing is selected for this study which is the NU205E.TVP2.C3 cylindrical roller bearing made from chrome steel. The roller bearing is designed to permit relative motion between two machine parts, usually between a rotating shaft and a fixed frame supporting the applied load. The load placed on the bearing is supported by the contact surface between the inner raceway and the rollers. The geometry of the bearing is shown in Table 1.

Table 1. Roller bearing geometry

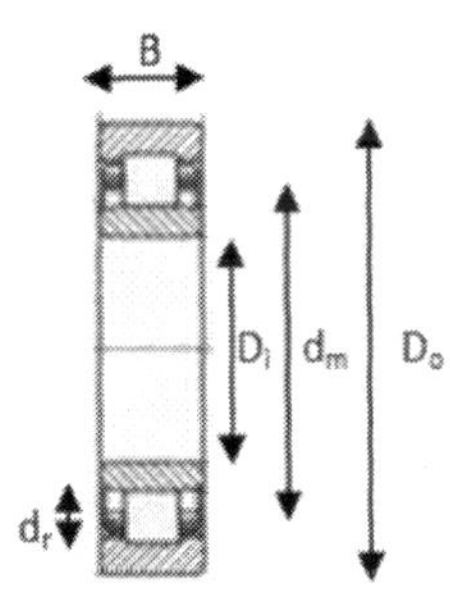

Number of roller	13
Roller diameter (dr)	7.5 mm
Inner bore diameter (Di)	25 mm
Outer ring outside diameter (Do)	52 mm
Pitch diameter (dm)	38.5 mm
Width (B)	15 mm
Weight	0.14 kg

A rotation system is designed and fabricated for this study. Figure 1 shows the experiment test rig which consists of bearings, motor, load, shaft coupling, speed controller and data acquisition system. The shaft is supported by two pillow blocks. The test bearing is fitted to one end of the shaft. During operation, the bearing inner ring rotated with the shaft while the outer ring remained stationary. The steel shaft is driven by an AC motor (Marelli Motor) and the speed controlled by speed controller

(Emerson). The shaft speed was set to 1500 rpm (25 Hz). The rotational speed of the shaft is measured using laser tachometer (Optel-Thevon). A load of 900 N is applied.

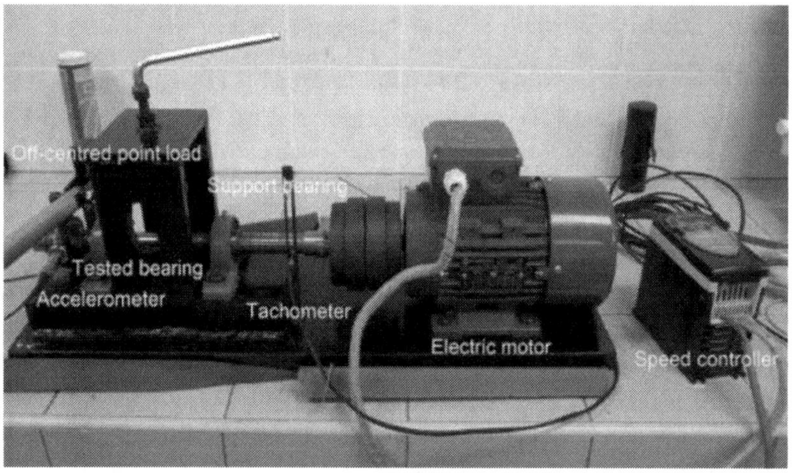

Figure 1. Experimental test rig

Vibration Measurement

The vibration level of roller bearings is measured by using accelerometer (DYTRAN) that is mounted horizontally on the bearing housing. The accelerometer calibration is performed at the beginning of the measurement. The shaft speed is monitored by tachometer probe. The signals from accelerometer and tachometer are processed with LMS Test.Xpress software and hardware with the sampling rate set at 5 kHz. The vibration data is analyzed in term of root mean square (RMS) value.

Lubricant Film Thickness Measurement

The bearings are tested for un-lubricated and lubricated conditions. For un-lubricated condition, the whole bearing is cleaned and submerged in ethanol for one hour. The intention is to remove all trace of lubricant for the bearing in the un-lubricated condition. For lubricated test, the SKF LGMT 2 0.4 grease is used in the experiment. The mineral base oil with Lithium thickener grease with viscosity of 110 mm^2/s at 40 °C and 100 °C at 11 mm^2/s is used. Surface roughness measurement is performed periodically in order to measure surface roughness for film thickness and lambda calculation. In order to do that, the lubricated bearing is removed every 30 minutes and cleaned for the characterization. For un-lubricated bearing, the bearing is removed more frequently as early damaged is expected. After characterization, the bearing is re-lubricated and remounted. The surface characterization is performed by using Infinite Focus Microscopy (IFM).

The lubrication condition is determined through minimum film thickness between the inner race and roller which present tribological contact in rolling bearing. The minimum film thickness, h_{min} for roller bearing is calculated using Hamrock-Dowson equation [16];

$$h_{min} = 3.63\, U^{0.68}\, G^{0.49}\, W^{-0.073}$$

where h_{min} is the min film thickness, U is speed parameter, G is material parameter and W is load parameter.

The relation between roughness and film thickness is defined by λ parameter as shown below [16], and can be used as an indicator of the lubrication regime.

$$\lambda = \frac{h_{min}}{\sqrt{R_{q1}^2 + R_{q2}^2}}$$

where R_q is a root mean square roughness height of two contact surfaces.

RESULTS AND DISCUSSIONS

The un-lubricated roller bearing shows signs of failure during the early stage of the experiment which produces noise and this gradually become worse until unacceptable noise and vibration level produced after 20 minutes. For the lubricated bearing it was considered damaged after producing high noise level after 110 hours (6600 minutes) of operation. The average roughness (R_a) measurement shows that the value is decrease initially as shown in Figure 2. Similar trend was also observed by [4, 17] where during this period, the continuous rolling contact between the roller and inner raceway causes plastic deformation for the two contacting surface asperities and reduces the roughness which is also known as polishing effect. For the un-lubricated bearing, the asperities polishing only took place for two minutes. It was explained that in the operation where an inadequate lubrication applied, the minimum film thickness is lower than surface asperities. Under un-lubricated condition where the severity of operation is increased, there is no lubrication film to support the load. The value started to increase after 7 minutes and 90 minutes for un-lubricated and lubricated condition respectively where the surface roughening took place until the bearing failure.

The minimum film thickness (h_{min}) calculated by using Hamrock- Dowson equation [16] is obtained at 0.36 μm. The relation between roughness and film thickness, (λ) is calculated to identify the lubrication regime of bearing running under grease lubrication. The assumption is that, the h_{min} is constant throughout the operation but the roughness change. The measured roughness during surface characterization is applied in the equation of measuring λ. The lubrication regime indicated by λ as function of time is shown in Figure 3. The figure shows that most of the time the bearing is operating in partial elasto-hydrodynamic (EHL) regime. The full film EHL regime is only achieved at the start of experiment. At the final of experiment, which at the period when the bearing shows a sign of failure, the boundary lubrication is reached.

In order to investigate the correlation between vibration and tribological phenomena, a vibration data for lubricated bearing are plotted against the λ parameter as shown in Figure 4. The low level of vibration is detected for lambda above 1.8 and unacceptable high level of vibration at lambda of 1.6 with growing level of intermediate value. Previous study [8] showed that for fully flooded lubrication, a λ=12 can be achieved initially and even for this lubrication indicated that vibration level started to increase when λ <6. Such results were based on calculated film thickness value of constant R_a. The results shown in figure 4 is important since the vibration value at the bearing clearly showed dependency on the λ value and critical lambda value of λ= 1.6 is the turning point where the vibration level begin to rise.

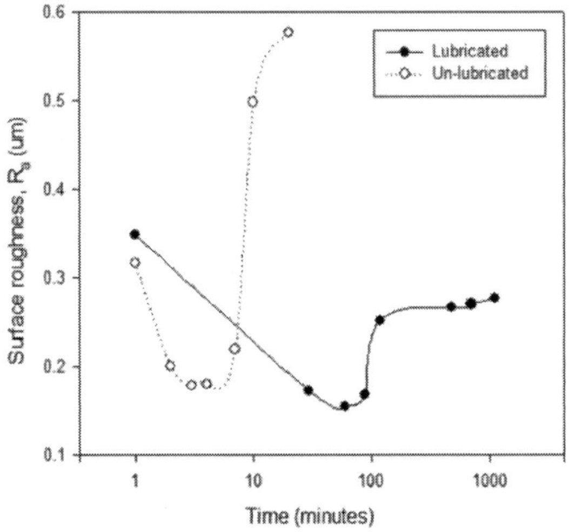

Figure 2 Surface roughness as a function of time

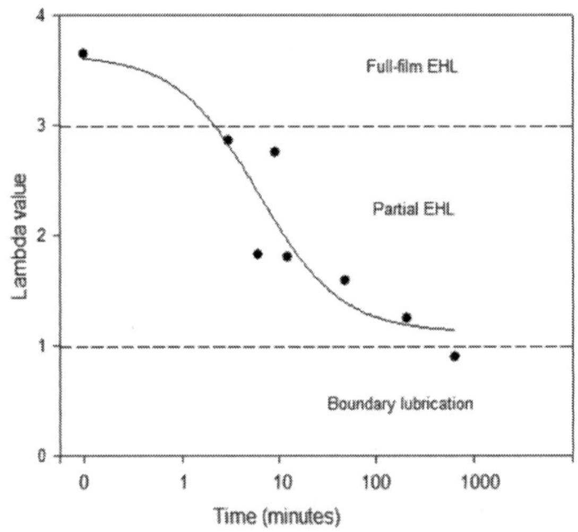

Figure 3. Lambda λ value as a function of time

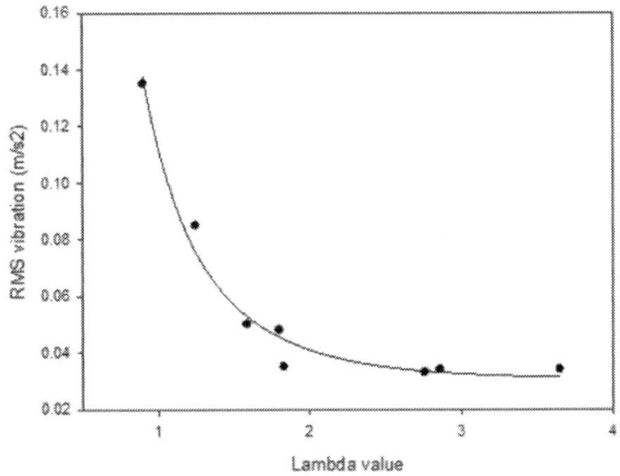

Figure 4 RMS vibration as a function of λ (film thickness/roughness)

CONCLUSION

The conclusions of this study are as follows:

1. The grease lubricated roller bearing has proved to be able to delay the surface degradation and failure. With un-lubricated condition, the surface roughening is accelerated and the vibration level is increased.

2. The lubrication plays an important role to ensure proper roller bearing operation. A high level of the vibration level is indicated at the smaller λ where the critical lambda value of λ = 1.6 is the turning point where the vibration level begin to rise.

ACKNOWLEDGEMENTS

The authors would like to acknowledge Universiti Sains Malaysia for providing the financial assistance under Short-term grant (304/PMEKANIK/60313052).

REFERENCES

[1] Norton, M. and Karczub, D. (2003). Fundamentals of Noise and Vibration Analysis for Engineers. Second Edition. Cambridge University Press.

[2] Patil, M.S., Mathew, J. and RajendraKumar, P.K. (2008). Bearing Signature Analysis as a Medium for Fault Detection: A Review. Journal of Tribology. 130:1 014001-7.

[3] Williams, T. (2001). Rolling element bearing diagnostics in run-to-failure lifetime testing. Mechanical Systems and Signal Processing. 15:5 979-993.

[4] Halme, J. and Anderson, P. (2009). Rolling contact fatigue and wear fundamentals for rolling bearing diagnostics. VTT Technical Research Centre of Finland. 224:J: 377-393.

[5] FAG Rolling bearing damage: Recognition of damage and bearing inspection. (1996). Publication no. WL 82 102/2 ED.

[6] Maru, M.M., Castillo, R.S. and Padovese, L.R. (2005). Detection of solid contamination in rolling bearing operation through mechanical signature analysis. In: Proceedings of 12th International Congress on Sound and Vibration. Portugal.

[7] Serrato, R., Maru, M.M., Padovese, L.R. (2005). Effect of lubricant oil viscosity and contamination on the mechanical signature of roller bearings. In: Proceedings of 12th International Congress on Sound and Vibration. Portugal.

[8] Serrato, R., Maru, M.M. and Padovese, L.R. (2007). Effect of lubricant viscosity grade on mechanical vibration of roller bearings. Tribology International. 40:8 1270-1275.

[9] Wunsch, F. (1992). Noise characteristic of lubricating greases used for anti-friction bearings. NLGI Spokesman 56:16–21.

[10] Boškoski, P., Petrovčič, J., Musizza, B. and Juričić, D. (2010) Detection of lubrication starved bearings in electrical motors by means of vibration analysis, Tribology International. 43:9 1683-1692

[11] Ebert, F.J. (2010). Fundamentals of Design and Technology of Rolling Element Bearings. Chinese Journal of Aeronautics, 23(1):123-136.

[12] Cann, P. M. E., Damiens, B. and Lubrecht, A. A. (2004). The transition between fully flooded and starved regimes in EHL. Tribology International 37: 859–864.

[13] Cann, P. M. E. (1999). Starved grease lubrication of rolling contacts. Tribology Transactions. 42: 867–873.

[14] Vengudusamy, B., Kuhn, M., Rankl, M. and Spallek, R. (2016). Film forming behavior of greases under starved and fully flooded EHL conditions, Tribology Transactions. 59:1 62-71

[15] Cen, H., Lugt, P.M and Morales-Espejel, G. (2014). Film thickness of mechanically worked lubricating grease at very low speeds, Tribology Transactions. 57:6 1066-1071.

[16] Hamrock, B.J., Schmid, S.R. and Jacobson, B.O. (1999). Fundamental of Machine Elements. Ohio. Mc Graw Hill.

[17] Akbarzadeh, S. and Khonsari, M.M. (2011). Experimental and theoretical investigation of running-in. Tribology International. 44:2 92-100

WHOLE BODY VIBRATION ANALYSIS OF BABY HAMMOCK

Ying Hao, Ko[1*], Chia Sin, Geh[1]
[1]Mechanical Engineering Department, School of Engineering and Technology (FOET)
Tunku Abdul Rahman University College,
*Email: koyh@tarc.edu.my
Phone: +60341450123 (Ext: 3544)

ABSTRACT

Studies have been carried on the effect of rocking on a baby and concluded that baby sleeps easier while being rocked. In Malaysia, as in many Southeast Asian Countries, it is common to put babies to sleep in a baby hammock. The vertical rocking motion generated by baby hammock has exposed babies to whole-body vibration (WBV). It has been shown by ISO2631 (1997) that WBV may lead the discomfort and adverse effect on health. Standards have been set by ISO 2631 (1997) concerning the WBV for people in a recumbent position and consider weighted vibrations of more than 2 m/s^2 to be extremely uncomfortable. However, standards concerning the allowable amount vibrations a baby in a baby hammock can safety endure are currently lacking. WBV analysis of the baby hammock with the weight ranged from 3kg to 14kg is conducted. For each measurement, four conditions are considered: manual rocking, auto rocking with low, medium and high speed. In this study, average root-mean-square values for the acceleration were found to be at a maximum of 2.46 m/s^2, and to be above the extremely uncomfortable level. This study develops a baseline exposure time for the baby hammock before it reaches the safety values of exposure action value (EAV) and exposure limit value (ELV) set by ISO 2631(1997).

Keywords: whole body vibration, baby hammock, vibration analysis, human comfort level

INTRODUCTION

It is very common to put babies into sleep by employing a gentle rocking motion because the rhythmic movement makes the babies felt like they were in the mother's womb ([1]-[3]). Hence, babies felt safe and comfortable while sleeping under the gentle rocking motion [2].

In Malaysia, as in many Southeast Asian Countries, it is common to put babies to sleep in a baby hammock during nap. Traditional baby hammock had been modernized into electric motor controlled baby hammock. Instead of manually rocking the baby hammock, the electric motor provides continuous rocking motion. The vertical rocking motion generated by baby hammock has exposed babies to whole-body vibration (WBV). Babies sleep at baby hammock are exposed to the possible effect from whole-body vibration. One of the effects is excessive vibration will cause negative effect on babies' brain development ([4]-[5]). Whole-body vibration depends on the magnitude, frequency and time of exposure. The longer the time exposure in baby hammock, the higher possibility of long-term health effects. Furthermore, high amplitude of the vibration generated by the electric motor controlled baby hammock might bring negative effect to babies' brain development [4]-[5]. According to Suima [6], babies fall asleep easily with gentle rocking at an amplitude of 60 mm and frequency of 0.55 Hz. This is because 0.55 Hz is similar to the mother's heart rate 67 bpm.

Whole-body vibration measurement has been done on baby car seat and prenatal incubator ([7]-[8]). However, there is no research has been carried out for whole-body vibration on baby hammock. Therefore, whole-body vibration analysis should be done on baby hammock. Babies usually take nap in baby hammock more than in a car seat. Hence, the time exposure of babies in baby hammock is more than the time exposure in car seat and in incubator. There are different range of sleeping time for different age of babies. The average nap time for a new-born baby to 2 years old is in between 2 hours to 9 hours [9].

However, there is no research has been done on whole-body vibration analysis of baby hammock. Hence, this paper aimed to analyse the whole-body vibration analysis of baby hammock based on International Standards Organization (ISO) 2631 [10] standards to evaluate the risk of whole-body vibration of baby hammock.

METHODOLOGY

Measurement of One-Third Octave Weighted RMS Acceleration

Figure 1a shows the baby hammock that was chosen to be the example of this project's analysis. It consists of timer and speed controller motor, a set of springs (7 pieces), a hanger and an adapter.

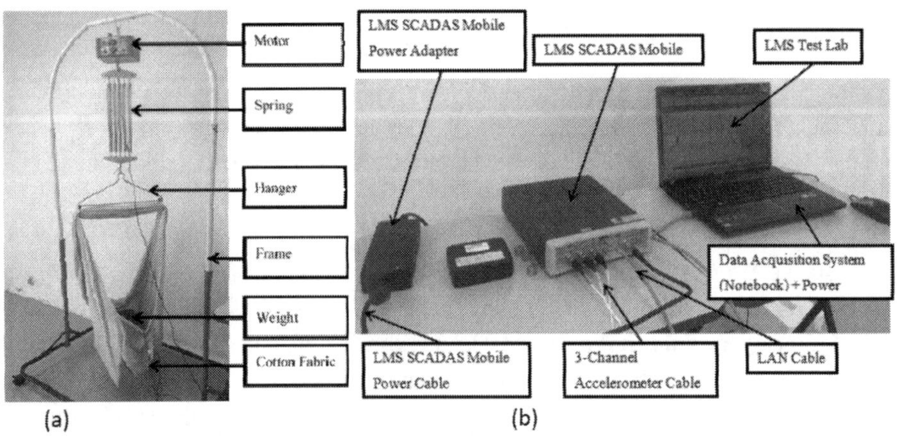

Figure 1. One-Third Octave weighted RMS acceleration Measurement setup

Figure 1b shows the devices and software used in this measurement. This includes a tri-axis accelerometer (DYTRAN, Type 3263A2), the analyzer (LMS SCADAS Mobile, Siemens) and software (LMS Test Lab). One-third octave frequency analysis were carried out in the baby hammock to determine the weighted RMS acceleration based on [10]. Whole-Body Vibration analysis were carried out for two types of rocking which are Automatic rocking and manual rocking. Automatic rocking consists of three different speeds: low, medium and high speed. The baby hammock were carried different loads weight from 3 kg to 14 kg to represent the babies' weight from new-born to 2 years old (No baby subject will involve). This is due to measurements were measured using triaxial accelerometer mounted on the weight, the output from the triaxial accelerometer are frequencies and amplitudes, These parameters are sufficient to calculate one-third octave weighted RMS acceleration and estimate the human sensitivity to vibrations for comfort (Table 1). The data obtained from the tri-axes accelerometer were stored and analyzed using LMS Test Lab where x-direction is referred to vertical recumbent direction while y and z direction are referred to horizontal recumbent direction. Then, one-third octave frequency analysis in the range of 0.63 Hz to 160 Hz were evaluated.

Table 1 : Human Sensitivity to Vibrations for Comfort [10].

Weighted RMS acceleration	Human sensitivity
Less than 0.315 m/s^2	not uncomfortable
0.315 m/s^2 to 0.63 m/s^2	a little uncomfortable
0.5 m/s^2 to 1 m/s^2	fairly uncomfortable
1.25 m/s^2 to 2.5 m/s^2	very uncomfortable
Greater than 2 m/s^2	extremely uncomfortable

Calculation of Overall Weighted RMS Acceleration

The overall weighted RMS acceleration for three directions: x, y and z-direction will calculate as stated in ISO 2631-1 (1997) :

$$a_{wx} = \sqrt{\sum_i (W_i a_i)^2} \qquad a_{wy} = \sqrt{\sum_j (W_j a_j)^2} \qquad a_{wz} = \sqrt{\sum_k (W_k a_k)^2} \quad \text{(Eq1)}$$

In these equations, the a_{wx}, a_{wy}, and a_{wz} represent the overall weighted RMS accelerations at i^{th}, j^{th} and k^{th} frequency for x, y and z-direction respectively; whereas W_i, W_j and W_k, are the weighting factors at i^{th}, j^{th} and k^{th} frequency shown in ISO 2631-1, 1997 and a_i, a_j and a_k are the RMS acceleration for the ith, j^{th} and k^{th} in x, y and z-direction [10].

Calculation of Total Vibration Value of Weighted RMS Acceleration

Total vibration value of weighted RMS acceleration is the combination of overall weighted RMS acceleration in three orthogonal directions: x, y and x-direction:

$$a_v = \sqrt{k^2_x a^2_{wx} + k^2_y a^2_{wy} + k^2_z a^2_{wz}} \quad \text{(Eq.2)}$$

a_v is the total vibration value of weighted RMS acceleration; a_{wx}, a_{wy}, and a_{wz} are the weighted RMS accelerations with respect to the orthogonal axes x, y, z, respectively and k_x, k_y and k_z are multiplying factors [10]. For a laying human, multiplying factor, k = 1 on both horizontal and vertical axes [10].

Calculation of Time to Reach the EAV or ELV

In order to prevent the long-term health effect or death risk from the vibration transmitted to body, one should need to know the safest exposure time to vibration. Hence, the time needed to reach the exposure action value (EAV) and exposure limit values (ELV) with a given total vibration value of weighted RMS acceleration can be calculated using Equation:

$$t_r = \left(\frac{EAV\ or\ ELV}{a_v}\right)^2 x\ 8 \quad \text{(Eq.3)}$$

Where t_r is the time need to reach the EAV or ELV and a_v is the total vibration value of weighted RMS acceleration.

RESULTS AND DISCUSSION

Total Vibration Value of Weighted RMS Acceleration for Comfort Level

As can be seen from Table 2, the total vibration values for weights from 3 kg to 14 kg subjected to automatic rocking with medium speed, high speed and manual rocking were mostly under the category of very uncomfortable. However, the total vibration values for weights from 3 kg to 14 kg subjected to automatic rocking with low speed were considered fairly uncomfortable for babies. And unfortunately, the total vibration value for weight 3 kg subjected to automatic rocking with high speed is 2.4581 m/s^2 which is extremely uncomfortable to babies. Hence, babies with lower weight is not recommended to use baby hammock subjected to automatic rocking with high speed.

Table 2: Total Vibration Value of Overall Weighted RMS Acceleration.

Springs	Weight (kg)	Whole-Body Vibration Acceleration (m/s^2)			
		Automatic Rocking			Manual Rocking
		Low Speed	Medium Speed	High Speed	
2	3	0.9548	1.6922	2.4581	1.5569

3	4	0.8806	1.5775	1.8219	1.3718
3	5	0.8709	1.5250	1.6902	1.1303
4	6	0.8871	1.3303	1.7647	1.3036
4	7	0.8442	1.2307	1.5367	0.9885
5	8	0.7688	1.0860	1.3930	1.3180
5	9	0.6546	1.0520	1.3463	1.0805
6	10	0.7279	1.0520	1.0319	0.8981
6	11	0.7222	0.8283	0.9780	0.6909
7	12	0.7038	1.3234	1.7549	1.2507
7	13	0.6902	1.3013	1.5924	0.9961
7	14	0.5825	0.8575	1.0365	0.9449

The data from Table 2 were also used to calculate the time to reach the EAV or ELV using the Equation 3. From Table 3, it is noted that baby hammock is safe to use up to 5 hours 53 minutes for weight 14 kg subjected to automatic rocking with low speed in order to stay below the EAV. However, it is only limited to 19 minutes for weight 3 kg subjected to automatic rocking with high speed. Hence, it can be concluded that the heavier the babies, the longer the time needed to reach the EAV and ELV for both types of vertical rocking motion. Therefore, it is recommended that heavier babies are more suitable to use the baby hammock.

Table 3: Time to reach EAV and ELV

Springs	Weight (kg)	Time to Reach EAV and ELV											
		Automatic Rocking									Manual Rocking		
		Low Speed			Medium Speed			High Speed					
		<EAV	EAV	ELV	<EAV	EAV	ELV	<EAV	EAV	ELV	<EAV	EAV	ELV
2	3	2 hr 11 min	> 2 hr 11 min	> 11 hr 36 min	41 min	> 41 min	> 3 hr 41 min	19 min	> 19 min	> 1 hr 45 min	49 min	> 49 min	> 4 hr 21 min
3	4	2 hr 34 min	> 2 hr 34 min	> 13 hr 38 min	48 min	> 48 min	> 4 hr 15 min	36 min	> 36 min	> 3 hr 11 min	1 hr 03 min	> 1 hr 03 min	> 5 hr 37 min
3	5	2 hr 38 min	> 2 hr 38 min	> 13 hr 56 min	51 min	> 51 min	> 4 hr 32 min	42 min	> 42 min	> 3 hr 42 min	1 hr 33 min	> 1 hr 33 min	> 8 hr 16 min
4	6	2 hr 32 min	> 2 hr 32 min	> 13 hr 26 min	1 hr 07 min	> 1 hr 07 min	> 5 hr 58 min	38 min	> 38 min	> 3 hr 23 min	1 hr 10 min	> 1 hr 10 min	> 6 hr 13 min
4	7	2 hr 48 min	> 2 hr 48 min	> 14 hr 50 min	1 hr 19 min	> 1 hr 19 min	< 6 hr 59 min	50 min	> 50 min	> 4 hr 28 min	2 hr 02 min	> 2 hr 02 min	> 10 hr 49 min
5	8	3 hr 23 min	> 3 hr 23 min	> 17 hr 54 min	1 hr 41 min	> 1 hr 41 min	> 8 hr 58 min	1 hr 01 min	> 1 hr 01 min	> 5 hr 27 min	1 hr 09 min	> 1 hr 09 min	> 6 hr 05 min
5	9	4 hr 40 min	> 4 hr 40 min	> 24 hrs	1 hr 48 min	> 1 hr 48 min	> 9 hr 33 min	1 hr 06 min	> 1 hr 06 min	> 4 hr 19 min	1 hr 42 min	> 1 hr 42 min	> 9 hr 03 min
6	10	3 hr 46 min	> 3 hr 46 min	> 19 hr 58 min	1 hr 48 min	> 1 hr 48 min	> 9 hr 33 min	1 hr 52 min	> 1 hr 52 min	> 9 hr 56 min	2 hr 28 min	> 2 hr 28 min	> 13 hr 07 min
6	11	3 hr 50 min	> 3 hr 50 min	> 20 hr 17 min	2 hr 54 min	> 2 hr 54 min	> 15 hr 25 min	2 hr 05 min	> 2 hr 05 min	> 11 hr 03 min	4 hr 11 min	> 4 hr 11 min	> 22 hr 09 min
7	12	4 hr 02 min	> 4 hr 02 min	> 21 hr 21 min	1 hr 08 min	> 1 hr 08 min	> 6 hr 02 min	38 min	> 38 min	> 3 hr 26 min	1 hr 16 min	> 1 hr 16 min	> 6 hr 45 min
7	13	4 hr 11 min	> 4 hr 11 min	> 22 hr 12 min	1 hr 10 min	> 1 hr 10 min	> 6 hr 14 min	47 min	> 47 min	> 4 hr 10 min	2 hr	> 2 hr	> 10 hr 39 min
7	14	5 hr 53 min	> 5 hr 53 min	> 24 hrs	2 hr 43 min	> 2 hr 43 min	> 14 hr 23 min	1 hr 51 min	> 1 hr 51 min	> 9 hr 50 min	2 hr 14 min	> 2 hr 14 min	> 11 hr 50 min

CONCLUSION

Based on the analysis, the comfort level of using the baby hammock based on ISO-2631-1 (1997) at different weights ranged from 3 kg to 14 kg (represents the babies from new-born to 2 years old) had been determined. From the results, the lowest total vibration value is 0.5825 m/s^2 (14 kg subjected to automatic rocking with low speed) which is considered as fairly uncomfortable. While the highest total vibration value is 2.4581 m/s^2 (3 kg subjected automatic rocking with high speed) which is considered as extremely uncomfortable. Furthermore, the time to reach the Exposure Action Value (EAV) and Exposure Limit Value (ELV) had been calculated based on the results from the total vibration values of overall weighted RMS acceleration. From the results, baby hammock is safe to use up to 5 hours 53 minutes for weight 14 kg subjected to automatic rocking with low speed. However, it is only limited to 19 minutes for weight 3 kg subjected to automatic rocking with high speed. Hence, it can be concluded that the heavier the babies, the longer the time needed to reach the EAV and ELV for both types of vertical rocking motion. As conclusion, a baseline exposure time for the baby hammock before it reaches its safety values has been developed.

ACKNOWLEDGEMENTS

The authors would like to acknowledge Tunku Abdul Rahman University College for the laboratory facilities and technical support for this project.

REFERENCES

[1] Pederson, D. R., Champagne, L. & Pederson, L. L. (1969). Relative Soothing Effects of Vertical and Horizontal Rocking. ERIC Number: ED046504.

[2] Bayer, L., Constantinescu, I., Perrig, S., Vienne, J., Vidal, P. P., Mühlethaler, M. & Schwartz, S. (2011). Rocking synchronizes brain waves during a short nap. Current Biology, 21(12), R461-R462.

[3] Barnard, K. E. (1972). The effect of stimulation on the duration and amount of sleep and wakefulness in the premature infant. PhD Thesis, University of Washington.

[4] Lu, C. H., Luo, C. H., Chen, Y. J., & Yeh, T. F. (1997). An automatic swinging instrument for better neonatal growing environment. Review of scientific instruments, 68(8), 3192-3196.

[5] Bang, S., Lam, R. & LoCicero, N. (2011). Rock Me Baby: The Automatic Baby Rocker. Project for San Jose State University, Department of Mechanical and Aerospace Engineering.

[6] Suima (2009). Suima: The Japanese Robocrib of the Future/Past. (http://daddytypes.com/2009/02/03/suima_the_japanese_robocrib_of_the_future past.php). Accesssed on 2 January 2015

[7] Nilsson, M. (2005). Health risk aspects and comfort for infants in infant seats for cars (VTI notat). Linköping: VTI., VTI notat 24-2005. Retrieved from http://urn.kb.se/resolve?urn=urn:nbn:se:vti:diva-1546

[8] Giacomin, J., Gallo, S. 2003. In-vehicle Vibration Study of Child Safety Seats. Ergonomics, 2003, vol 46, no 15, 1500-1512

[9] Denise Fields and Ari Brown, MD. 2008. National Sleep Foundation, Baby 411 and Toddler 411.

[10] ISO 2631-1:1997, Mechanical vibration and shock – Evaluation of human exposure to whole-body vibration – Part 1: General requirements.

ATTENUATION OF MOTORCYCLE HANDLE VIBRATION USING DYNAMIC VIBRATION ABSORBER

Muhammad Iyad Al-Maliki Saifudin, Nabil Mohamad Usamah and Zaidi Mohd Ripin

TheVibrationLab,
School of Mechanical Engineering,
Universiti Sains Malaysia,
14300 Nibong Tebal, SPS,
Pulau Pinang, Malaysia
*Email: mezaidi@usm.my, thevibrationlab@gmail.com
Phone: +6045996368; Fax: +6045996912

ABSTRACT

Motorcycle riders are exposed to hand-transmitted vibration of the hand-arm system due to the vibration of the handle and extended exposure can result in numbness and trembling. One feasible solution to attenuate the handle vibration is by using a dynamic vibration absorber (DVA). In this work a DVA is designed and mounted on the motorcycle handle in order to reduce the vibration at the handle by transferring the vibration from the primary system handle to the secondary mass. Removal of elastomeric material at the DVA mounting locations, symmetry of secondary mass and the direction of DVA attachment influence the vibration absorption. A series of tests conducted show that the vibration on the handle is mainly induced by the engine and there is additional source of vibration from the road surface roughness. Installation of DVA at different locations on the handle resulted in various attenuation levels at different speed in the x and z directions. The attenuation level is between 59-68 % in the biodynamic x-directions for speed at 30-50 kmh^{-1}.

Keywords: Motorcycle Handle, Vibration Absorber, Hand-Arm Vibration, and Vibration Analysis.

INTRODUCTION

Motorcycles are one of the major transportation vehicle used in the world because they are cheap with low fuel consumption and easy to handle [1]. The engine vibration and road conditions can result in excessive vibration which will reduce the user comfort [2][3]. The excessive exposure of vibration to the user can cause vascular disorder, muscle disorder, neurological disorder and bone and joint disorder which can be called hand arm vibration (HAVs) [4]. Studies showed that more than 50 % of respondents suffer from discomfort at the hands and the arm and more than 50 % of non-occupational motorcyclist experienced discomfort at their low back, neck, shoulders, elbows, and upper back. Meanwhile occupational motorcyclists experienced greater discomfort [2][5]. 4.2 % of motorcyclist police officer suffered from finger blanching [6]. Standard human vibration exposure limit set by the European Directive (European Directive/2002/44/EC) can be used to achieve the acceptable level to reduce the prevalence of hand arm vibration syndrome. The daily exposure action value (EAV) required users to reduce the any vibration exposure above than 2.5 ms^{-2} while the exposure limit value (ELV) is 5.0

ms^{-2}. From previous research, the frequency of interest of a motorcycle vibration is in the range of 50 Hz to 300 Hz while the vibration magnitude on the motorcycle handlebar is between 2.2 ms^{-2} to 9.77 ms^{-2} [2][6][7][8]. The high variance of the exposed vibration magnitude of different respondents are related to several uncontrolled parameters [2]. Some of the uncontrolled parameters are the user handlebar griping force [9][10] and the area of contact between the hand to the vibrating components [11]. Numerous studies have been done to overcome the hand arm vibrations. One technique is with the use of vibration absorber [12][13]. It is a tunable secondary mass with spring and damping properties with natural frequency similar to the excitation force frequency. Examples of application are in electrical grass trimmer [14], a motorcycle handle [7], machine [15] and hand blender [16]. Most studies in the past were mainly done by measuring handle vibration with the motorcycle stationary on the dynamometer [7] or at relatively low speed of up to 10 kmh^{-1} [2] or at a mean velocity of 40 kmh^{-1} [8]. This paper covers the development of dynamic vibration absorber for use on motorcycle handlebar which include on the road test of speed range at 30-50 kmh^{-1}.

METHODOLOGY

Modal analysis of motorcycle handlebar

Figure 1 shows the measurement nodes located on the motorcycle handlebar labelled with stickers. Modal analysis was carried out with the motorcycle handlebar suspended using rubber chords to represent the free-free condition. A small light weight accelerometer (Dytran 3055B2T) was used to measure the vibration response in the x-axis while an impact hammer (Krisler 9724A5000) was used to excite the motorcycle handlebar. Both are connected to the 8-channel LMS SCADAS mobile. The accelerometer is calibrated and placed on the measured nodes using the roving accelerometer techniques. The signals from the test were recorded to derive the FRF (Frequency Response Function) for each measurement nodes, post processed and displayed on LMS Impact Testing Rev. 8B Software.

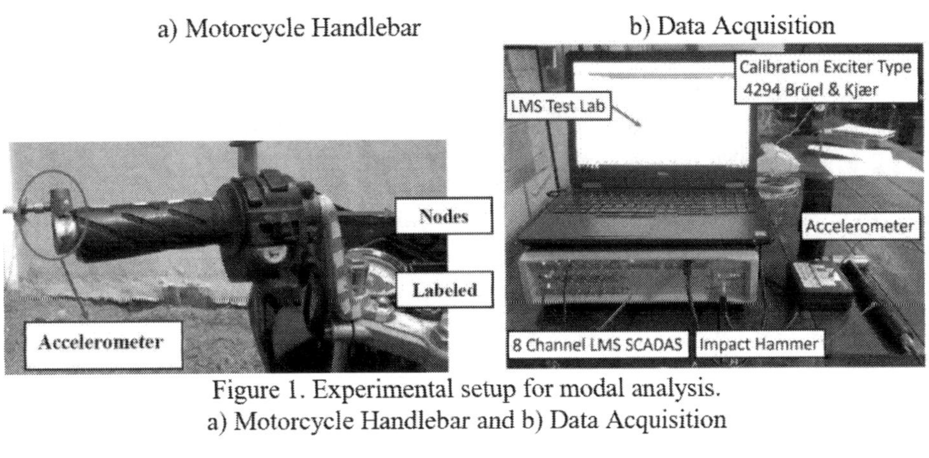

Figure 1. Experimental setup for modal analysis.
a) Motorcycle Handlebar and b) Data Acquisition

Vibration analysis of Suzuki Belang 150R handlebar on the Road Test

In order to determine the operating frequency, vibration level and their dominant axis, vibration measurement is carried out on the motorcycle handle. The measurement instruments include one miniature triaxle accelerometer (Dytran, 3023M20), calibrator (B&K, 4294), FFT analyzer (LMS Scadas Mobile), post-processing software (LMS Test Lab), tachometer and a portable computer. The measurement is carried out over a distance of 110 meters straight-flat asphalt road surface. Accelerometers are located on the handlebar, the rider left hand and the motorcycle fork. The accelerometers are connected to the LMS Scadas Mobile running on LMS Test Lab and a portable computer for the data acquisition system. The test is carried out at three different constant speed ranging from 30 kmh^{-1} to 50 kmh^{-1} at 10 kmh^{-1} interval. The vibration level is also measured at the steel parts of the handlebar near the left handgrip. The direction defined is as in the biodynamic axes. The handgrips are held by the rider. Similar experiment is carried out with the attachment of the dynamic vibration absorber to measure and evaluate the effectiveness of the DVA.

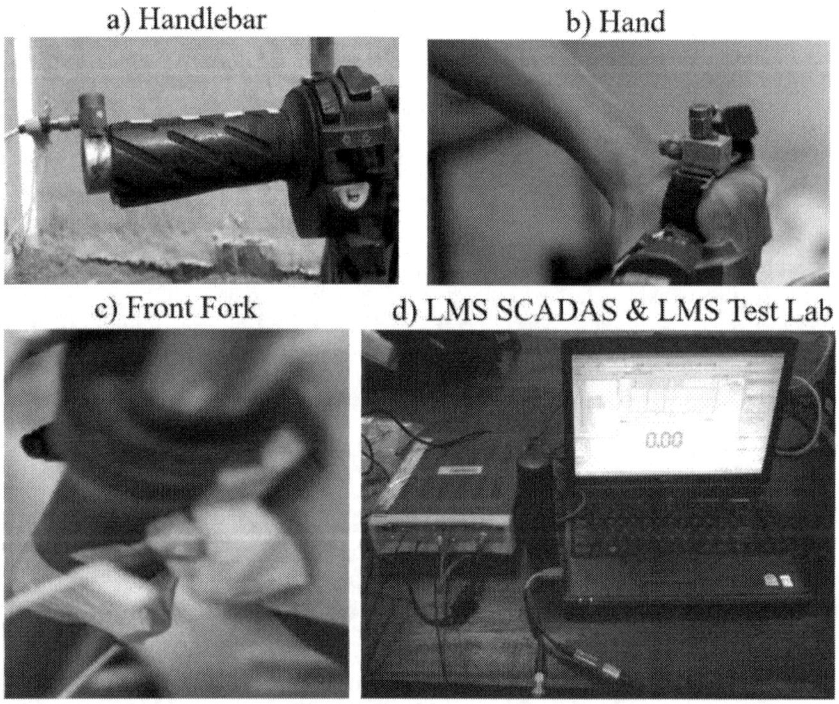

Figure 2. Schematic diagram for the vibration analysis on road test.
a) Handlebar accelerometer b) Hand Accelerometer
c) Front Fork Accelerometer and d) LMS SCADAS & LMS Test Lab

https://doi.org/10.1051/matecconf/201821701006

RESULTS AND DISCUSSION

Modal Analysis of Motorcycle Handlebar

The handlebar dimension of the motorcycles was measured and this is then used in the drawing of the "node" & "line" functions of the LMS Test Lab to visualize and analyze the vibration for modal analysis. Figure 3 shows the FRF of the handle in x and z axes under impact testing. For the x-axis, Mode 2 is the only mode with high participation of 62 % at 131 Hz. While for the z-axis, Mode 1 and Mode 6 with 40 % and 23 % mode participation at 60 Hz and 297 Hz respectively. All vibration modes are below 400 Hz and within the frequency range significant to human arm vibration.

a) x-axis b) z-axis

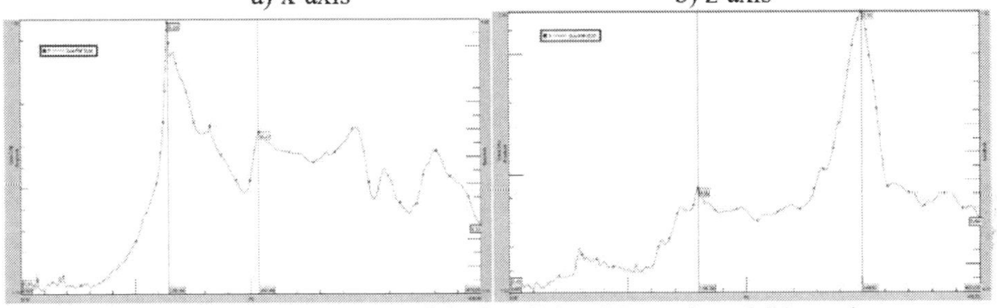

Figure 3. FRF Spectrum of the handlebar (a) x-axis and (b) z-axis

Motorcycle handlebar Response Spectrum on the Road Test

The RMS acceleration for a series of road tests are listed in Table 1 based on the acceleration of the handlebar at node 1. The accelerations are highly affected by the speed. The dominant direction is in the x-axis and the z-axis acceleration is always lower and does not follow the trend of the acceleration in x-axis. The response spectra for the RMS accelerations in the x-axis and z-axis are plotted in figure 4. Both axes exhibited frequencies clustering around a fixed frequency at 160 Hz. The relatively constant peak frequency in this case is basically speed independent, which indicated the possibility of resonance within this speed range.

a) x-axis b) z-axis

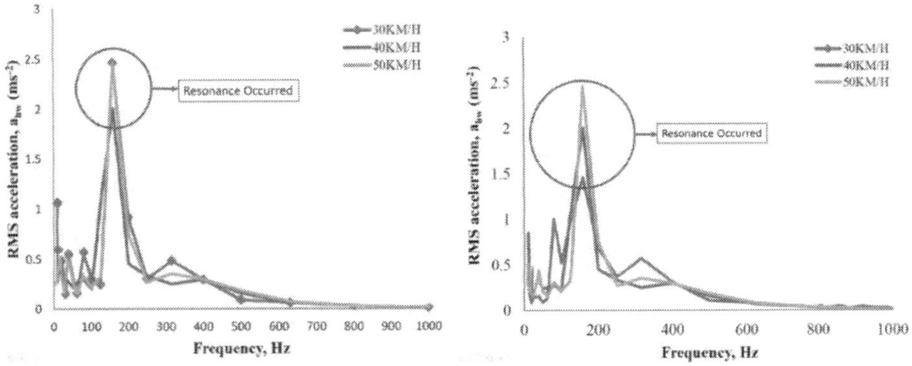

Figure 4. RMS Acceleration Response Spectrum (a) x-axis and (b) z-axis

Table 1. RMS acceleration for handle

Input Speed (kmh^{-1})	Acceleration (ms^{-2})		Peak Frequency (Hz)
	x	z	
30	2.487	1.470	160
40	2.218	2.023	160
50	2.482	1.751	160

Vibration Attenuation of motorcycle handlebar using added mass

Figure 5 below shows the comparison of vibration level on the motorcycle handlebar during road test evaluation before and after the installation of the added mass. The higest vibration peak of the x-axis and z-axis at 207.5 Hz shifted to 200.0 Hz, with the peak vibration attenuation is 59 % and 50 % from 3.32 ms^{-2} to 1.35 ms^{-2} and 3.29 ms^{-2} to 1.65 ms^{-2} respectively.

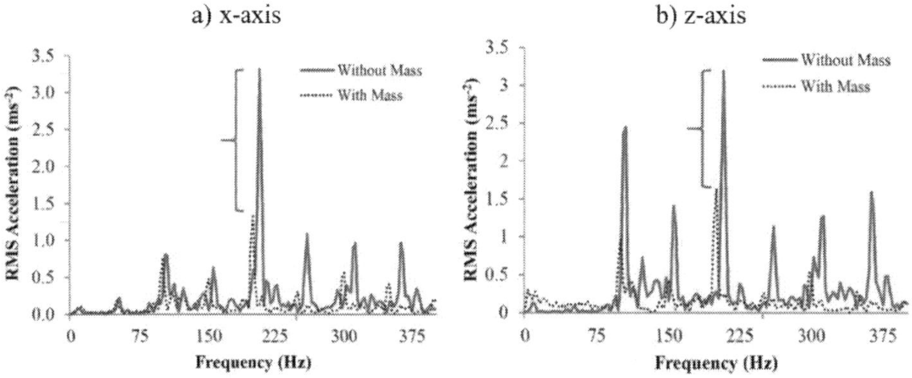

Figure 5. The effect of added mass on RMS acceleration (a) x-axis and (b) z-axis

Vibration Attenuation of motorcycle handlebar using DVA

Figure 6 show the RMS acceleration level of the motorcycle handlebar during road test before and after the installation of DVA. Each response spectra varies with different input speed. Focussing on the x-axis with the tuned DVA frequency of 200 Hz, the attenuation ranges from 59% to 68%. Based on the highest vibration peak amplitude, the attenuation level for 30 kmh^{-1}, 40 kmh^{-1} and 50 kmh^{-1} are 59 %, 60 % and 68 % respectively. A lower vibration amplitudes are measured in the z-axis. With the tuned DVA ferquency of 200 Hz, the attenuation level is between 34 % to 47 %. For the full response spectrum, based on the highest vibration peak amplitude the attenuation level for each speed is shown in figure 7. It is interesting to note that the attenuation level is inreasing with the speed which indicate that the DVA is functioning well, particularly attenuating the vibration on the resonant frequency of the handlebar. The level of attenuation at 59-68 % is typical and comparable to the attenuation level achieved by using DVA elsewhere [7][14][15][16].

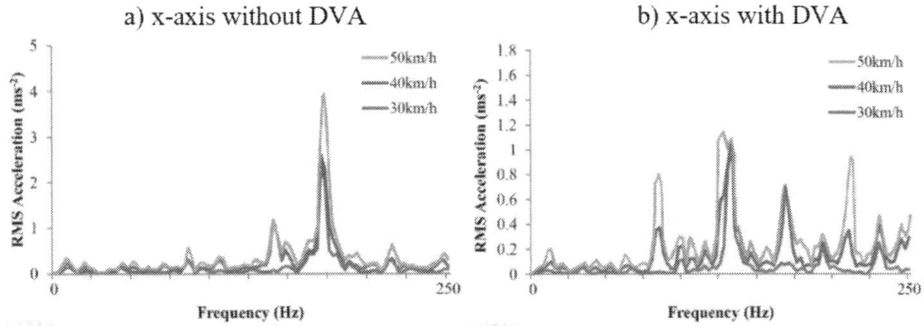

Figure 6. RMS vibration acceleration during road test with and without DVA (a) x-axis without DVA and (b) x-axis with DVA

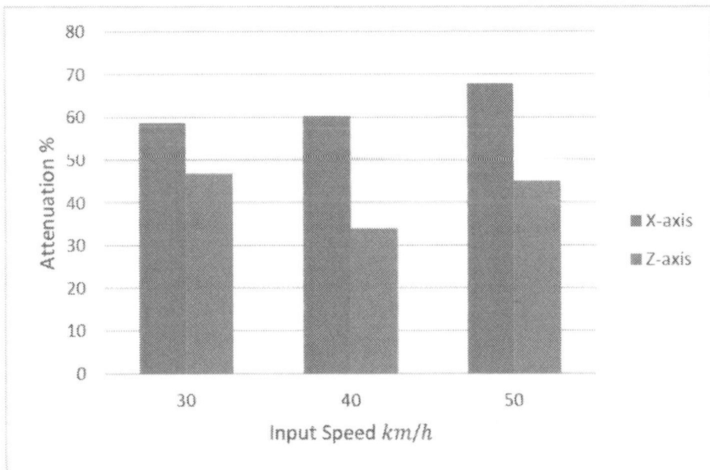

Figure 7. Vibration attenuation level at different input speed

CONCLUSION

A dynamic vibration absorber and added mass are deployed to attenuate the motorcycle handlebar vibration. The peak attenuation level of 59-68 % are achieved depending on the speed of 30-50 kmh^{-1}.

ACKNOWLEDGEMENTS

The authors would like to thank the Ministry of Higher Education and USM for providing the facilities and the research opportunity through Bridging Grant A/C 304/PMekanik/6316214

REFERENCES

[1] Anisa Holmes, "The Top 5 Cycling & Motorcycling Countries in the World - Dalia Research," *Dalia*, 2017. [Online]. Available: https://daliaresearch.com/blog-the-top-5-cycling-motorcycling-countries-in-the-world/. [Accessed: 27-Jun-2017].

[2] J. M. Noh, K. A. M. Rezali, A. As, and N. A. A. Jalil, "Transmission of Vibration from Motorcycle Handlebar to the Hand," *J. Soc. Automot. Eng. Malaysia*, vol. 1, no. 3, pp. 191–197, 2017.

[3] Y. Patel, J. Trivedi, S. Dadhaniya, B. Patel, A. J. Sheth, U. G. Student, and P. A. J. Sheth, "Design and Development of Tuned Vibration Absorber to Control Engine Body Vibration in Idling and Running Condition," *Int. J. Eng. Sci.*, vol. 11561, no. 5, pp. 11561–11563, 2017.

[4] M. J. Griffin, "Dose-effect relationships for hand-transmitted vibration.," *Handb. Hum. Vib.*, pp. 609–631, 1990.

[5] K. Karmegam, M. Y. Ismail, S. M. Sapuan, N. Ismail, M. T. Shamsul Bahri, S. Shuib, and P. Seetha, "A study on motorcyclist's riding discomfort in Malaysia," *Eng. e-Transaction*, vol. 4, no. 1, pp. 39–46, 2009.

[6] S. M. Mirbod, H. Yoshida, M. Jamali, K. Masamura, R. Inaba, and H. Iwata, "Assessment of hand-arm vibration exposure among traffic police motorcyclists," *Int. Arch. Occup. Environ. Health*, vol. 70, no. 1, pp. 22–28, 1997.

[7] A. Fasana and E. Giorcelli, "A vibration absorber for motorcycle handles," *Meccanica*, vol. 45, no. 1, pp. 79–88, 2010.

[8] L. M. Roseiro, M. A. Neto, A. M. Amaro, C. J. Alcobia, and M. F. Paulino, "Hand-arm and whole-body vibrations induced in cross motorcycle and bicycle drivers," *Int. J. Ind. Ergon.*, vol. 56, pp. 150–160, 2016.

[9] N. A. Mann and M. J. Griffin, "Effect of contact conditions on the mechanical impedance of the finger.," *Cent. Eur. J. Public Health*, vol. 4, no. 1, pp. 46–49, 1996.

[10] P. Marcotte, Y. Aldien, P.-É. Boileau, S. Rakheja, and J. Boutin, "Effect of handle size and hand–handle contact force on the biodynamic response of the hand–arm system under zh-axis vibration," *J. Sound Vib.*, vol. 283, no. 3–5, pp. 1071–1091, 2005.

[11] K. A. Md Rezali and M. J. Griffin, "Transmission of vibration through gloves: effects of material thickness," *Ergonomics*, vol. 60, no. 1, pp. 69–81, 2016.

[12] P. Watts, "On a method of reducing the rolling of ships at sea," *Trans. Inst. Nav. Archit.*, vol. 24, pp. 165–190, 1883.

[13] H. Frahm, "Device for damping vibrations of bodies." Google Patents, 18-Apr-1911.

[14] Y. H. Ko, O. L. Ean, and Z. M. Ripin, "The design and development of suspended handles for reducing hand-arm vibration in petrol driven grass trimmer," *Int. J. Ind. Ergon.*, vol. 41, no. 5, pp. 459–470, 2011.

[15] E. V Golysheva, V. I. Babitsky, and A. M. Veprik, "Vibration protection for an operator of a hand-held percussion machine," *J. Sound Vib.*, vol. 274, no. 1–2, pp. 351–367, 2004.

[16] A. J. Sheth, "Vibration control of a hand blender with the tuned vibration absorber," *Int. J. Adv. Res. Eng. Appl. Sci.*, vol. 4, no. 10, pp. 22–30, 2015.

SOUND AND VIBRATION DAMPING PROPERTIES OF NONWOVEN FLAX REINFORCED ACRYLIC BASED POLYESTER COMPOSITES

WMA Wan Mamat Ali[1*], L.E.Ooi[1], and Z. A Mohd. Ishak[2]

[1]TheVibrationLab,
School of Mechanical Engineering,
Universiti Sains Malaysia,
[2]Science and Engineering Research Centre,
Engineering Campus,
Universiti Sains Malaysia.
*Email: amri@usm.my
Phone: +6045996346

ABSTRACT

This paper presents the sound and vibration damping properties of nonwoven flax reinforced acrylic based polyester composites. This research attempt to solve the problem of synthetic fibre which caused environmental problems. The goal of is study to characterize the sound absorption and damping properties of the natural fibre at difference thickness and pressure. The nonwoven flax reinforced acrylic based polyester composites are produced by using moulding machines at different compression pressures that are 10, 20 and 30 bars. The thicknesses of samples used in this study are 2 mm and 3 mm. The measurement of sound absorption of the material is carried out using impedance tube. For the sample with 2 mm thickness, the result shows that the maximum sound absorption of 0.93 was observed at frequency of 853 Hz and pressure of 20 bars. Damping values are calculated and measure for each sample. The result shows that damping factor is obtained is 0.039 and 0.052 for thickness materials of 2mm and 3mm.

Keywords: nonwoven flax reinforced acrylic based polyester composites, vibration damping and sound absorption coefficient.

Introduction

Today, noise pollution has become a new pollution source that caused bad influence to the environment, human health and national economy. Researchers have carried out various noise control methods for investingating the issue. The mechanism of controlling noise included active control and passive control [1]. Recent years, researchers have focus on natural fibers such as kenaf, ramie and jute reinforced composites. The reason is good mechanical properties, light weight, environmentally friendly and biodegradable [2,3].

Multiple layered coir fiber absorbers involving air spaces and micro-perforated plate has been studied both through simulation and experiments [4-5]. The absorption performance of industrial waste tea leaf fibre has been studied by Ersoy and Kucu [6].

They found that the absorption coefficient of the fibres increases significantly when backed with a single woven cotton cloth. Three types of natural fibres which are ramie, flax and jute in a form of composites has been studied by Yang and Li [1]. It is found that the natural composites had satisfying noise reduction ability similar to that of the synthetic fibres. Noise control panels made from vegetable fibers consisting coconut, palm, sisal and açaí is proposed by Bastos et al. [7]. It was found that the performance of vegetable fibers could be greater in certain cases when compared to the synthetic acoustic foam materials. Subsequently, sound absorption properties of kapok fibers have been studied by Xiang et al. [8]. The authors found out that kapok fibres have similar acoustics performance with the combinations of commercial glass wool and cotton fibres although the bulk density of kapok fibre is lower. Putra et al. [9] studied sugarcane waste fibers and found that the sound absorption coefficient is comparable with those of commercial synthetic fibers. The review shows that there have been many studies performed on natural material based which take advantage of both natural materials as well as superior mechanical performance over metallic structures and also it has good sound absorbing and vibration damping properties [10]. These natural materials could essentially be grown for the purpose of manufacturing composites, in turn providing such benefits as being both biodegradable and recyclable.

The goal of this study is to explore the sound absorbing and damping properties of composites composed with natural materials and compare them over commonly used traditional composites, such as glass-reinforced composites.

Experimental Measurement

Material preparation

Natural fiber is capable of replacing glass fiber and synthetic fibers in automotive and construction industries, due to its lightweight and low costs [11], [12]. In addition, natural fiber such as kenaf, hem, flax, jute and sisal reinforced composites can reduce the weight and manufacturing cost, reducing the use of carbon dioxide (CO2) for manufacturing, recyclable and the main benefits are "green" and environmentally friendly. In this research, a non-woven flax material is used to produce non-woven flax reinforced acrylic based polyester composites. The material used is Acrylic Based Polyester resin (Acrodur) 950L with a real density of 1300 g/m^2.

Flax fiber material (non-woven flax) is cut according to the desired size, and the fiber section will be placed in air circular oven for 12 hours at 60 ±5 °C in order to ensure the material is fully dry and free from moisture. Next, the material is immersed into the liquid of Acrodur 950 L resin for 5 minutes and then dried for 30 minutes in the circular oven. The materials are been inserted into the hydraulic hot press machine at temperature of 210 ^{o}C. The materials are pressed using the hydraulic hot press machine at difference pressures that are at 10 bars, 20 bars and 30 bars. For this process, the time is set for six minutes for each press. Figure 1 shows the compression mold machine that used in this research. Finally, the materials are cut using the shear machine according to the shape and size required.

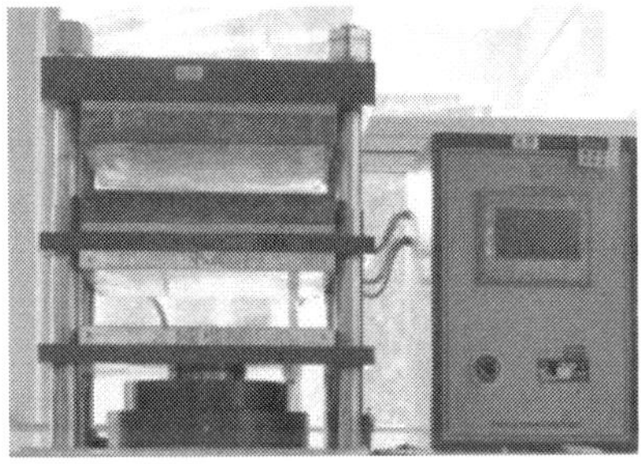

Figure 1: Compression Mould Machine

Sound absorption coefficient measurement

Sound absorption coefficient of each sample is measured using impedance tube, LMS Scadas Mobile, LMS software (LMS Sound Absorption Coefficient) and two units of microphones (BSWA MA211) as shown in Figure 2. The plunger is used to control the air gap. Microphones are calibrated each time before measurement. The samples used in this measurement are prepared in 2 mm and 3 mm as shown in figure 3. The loudspeaker was used as a sound source for the impedance tube that connected to one end as illustrated in Figure 2. The loudspeaker generates the broadband random sound waves and the sound waves propagating as plane waves in the tube hit the sample, get partially absorbed, and subsequently reflected. The acoustical properties of the test sample were tested in the frequency range of 100–1500 Hz. This system tests a sound absorptive material, processes the results, and reports the results in a graph of the absorption coefficient in various frequencies. Thus, the absorption coefficient of each sample was obtained. Table 1 list out the dimensions and pressure for different specimens used in this research.

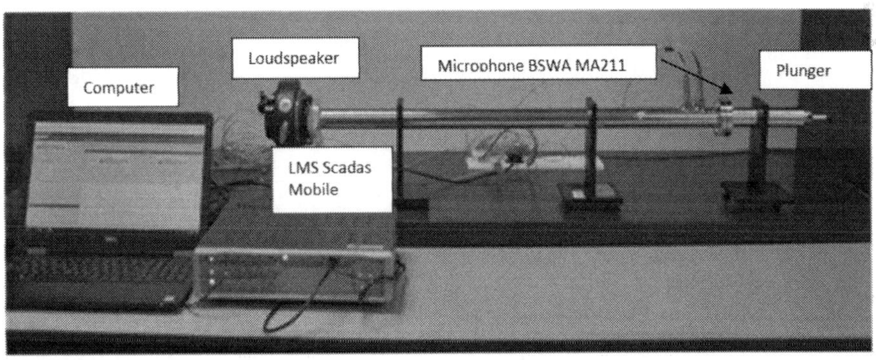

Figure 2: The setup experiment of the sound absorption coefficient.

Table 1: Dimension of sound test specimens

Pressure (bar)	Thickness (mm)	Diameter (mm)	Weight (grams)
10	2	34.8	1.87
20	2	34.8	1.82
30	2	34.8	1.77
10	3	34.8	1.82
20	3	34.8	1.84
30	3	34.8	2.02

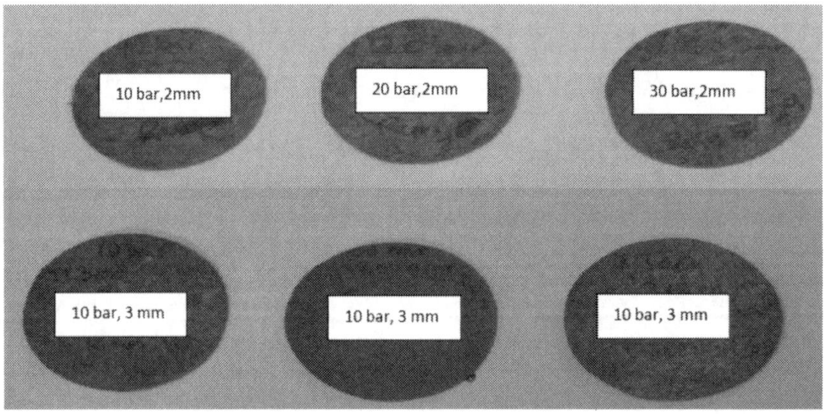

Figure 3: Sound Test Specimens of thickness of 2 mm and 3mm

Vibration damping factor measurement

Free vibration method is used to measure damping properties of each sample as shown in Figure 5. The specimens are cut according to the desired size of the different specimens as shown in Table 2 and Table 3. The specimen is mounted on cantilever beam structure and tied with a fixture. A Dytran accelerometer (model Model 3224A2) sensor with sensitivity of 2.89 mV/g is placed at the end of the specimen. Output signals are received by connecting the accelerometer sensor with Data Acquisition of LMS Scadas Mobile which is connected to Personal Computer (PC) with LMS Test Xpress software.

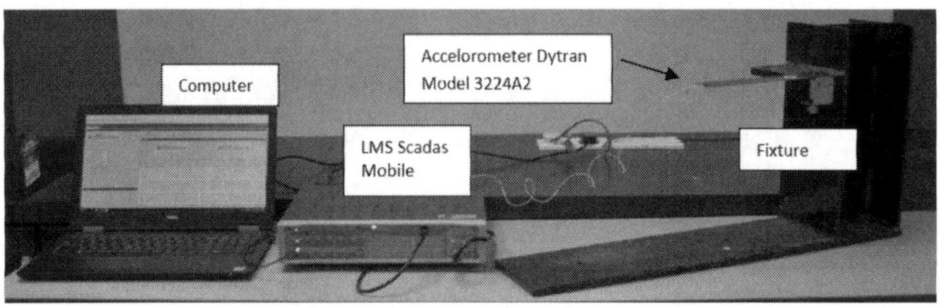

Figure 5: Vibration damping test set up

Table 2: Dimensions of vibration test specimens (2 mm and 3 mm)

Pressure (bar)	Thickness (mm)	Length (mm)	Width (grams)
10	2	210	35
20	2	210	35
30	2	210	35
10	3	210	35
20	3	210	35
30	3	210	35

Figure 6 shows the vibration response of the specimen over time. The damping factor is calculated by taking the highest value and the lowest value at the peak of the graph in the waveforms obtained. Then take some number of peak graphs for the damping factor calculation (ζ) and the values obtained are included in equations 1 and 2.

$$\delta = \frac{1}{\eta}\ln(\frac{y_0}{y_n}) \quad (1)$$

$$\zeta = \frac{1}{\sqrt{1+(\frac{2\pi}{\delta})^2}} \quad (2)$$

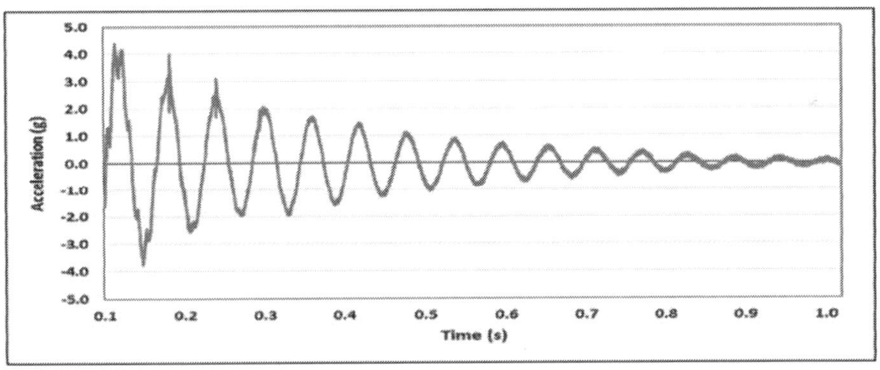

Figure 6: Free vibration response of the composite

Results and discussions.

Analysis of sound absorption coefficient measurement

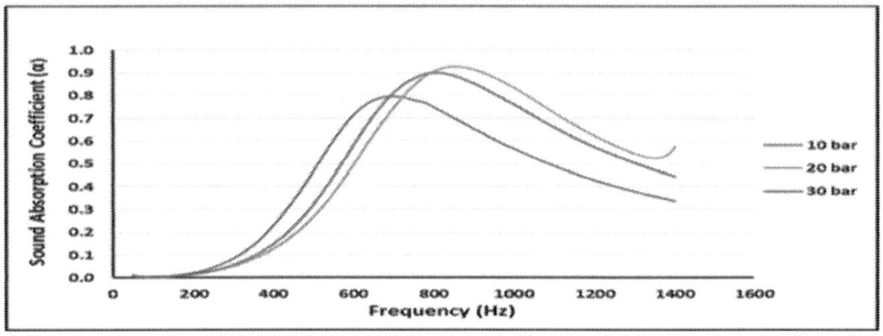

Figure 7: Sound absorption coefficient for specimen with thickness of 2 mm.

Figure 7 shows the results of sound absorption coefficient for specimen of thicknesses of 2 mm. In figure 7, the sound absorption coefficient increases with the frequency and reaches the maximum value and then begins to decrease as the frequency is increased. The maximum sound absorption coefficient for specimen at preesure 20 bar reach 0.93 at frequency of 853 Hz. For sample with pressure of 10 bars, sound absorption coefficient os 0.80 at frequency of 700 Hz. While for sample at 30 bars, it is observed that the sound absorption coefficient of 0.90 is recorded at frequency of 803 Hz.

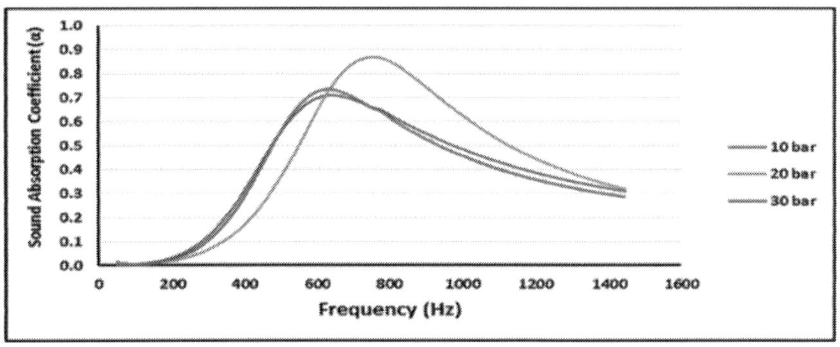

Figure 8: Sound absorption coefficient for specimen with thickeness of 3 mm.

Figure 8 shows the results of sound absorption coefficient for specimens with thicknesses of 3 mm. The figure shows sound absorption coefficient increases with the frequency and reaches the maximum values and begins to decrease as the frequency is increased. The maximum sound absorption coefficient of 0.87 is achieved at frequency of 760 Hz for sample with pressure of 20 bars. For sample with pressure of 10 bars, 0.71 of sound absorption coefficient was observed at frequency of 640 Hz. While for samples at 30 bars, it is observed that the sound absorption coefficient of 0.73 is observed at frequency of 803 Hz.

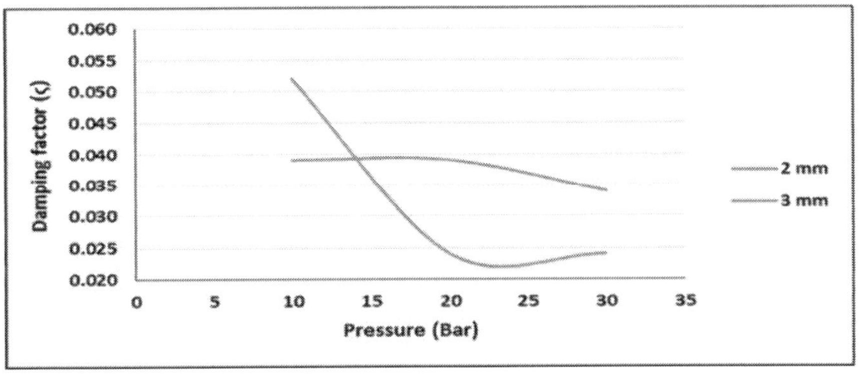

Figure 9: Damping ratio for specimen 2 mm and 3 mm under different compression pressure (bar)

Figure 9 shows the result of damping ratio for specimen of thickness of 2 mm and 3 mm for nonwoven flax reinforced acrylic based polyester composites that measured by LMS Test Xpress software. For specimen with thickness of 2 mm, the maximum damping factor is 0.039 at pressure of 10 bars and 20 bars. The damping factor slightly drops to 0.034 when the pressure is increased to 30 bars. For specimen with thickness of 3 mm, the maximum damping factor is 0.052 at pressure of 10 bars. The damping factor drop to its minimum point at 20 bars and increased to 0.024 at pressure of 30 bars. The measured SAC and damping factor in this paper obtained is archived similar trend compared to research conducted by Prabhakaran et al. [10] which produced the SAC and damping ratio of 0.45 and 0.014 respectively.

Conclusion

Nonwoven flax reinforced composite fibres have been shown to possess good sound absorption through the impedance tube measurement. The normal-incidence sound absorption coefficient of nonwoven flax fibre is found to have value of 0.93 at 853 Hz for specimen with thickness of 2 mm and pressure of 20 bars. Increasing the pressure can further enhance the sound absorption coefficient. These results suggest that the nonwoven flax reinforced composites could be viable candidate for applications which need of good sound and vibration properties.

Acknowledgement

The authors would like to gratefully acknowledge Universiti Sains Malaysia (USM) Short Term Research Grant [grant no. 60313028] for funding the project.

References

[1] W. Yang and Y. Li, "Sound absorption performance of natural fibers and their composites," *Sci. China Technol. Sci.*, vol. 55, no. 8, pp. 2278–2283, 2012.

[2] A. K. Mohanty, M. Misra, and L. T. Drzal, "Surface modifications of natural fibers and performance of the resulting biocomposites: an overview," *Compos. interfaces*, vol. 8, no. 5, pp. 313–343, 2001.

[3] H. Cheung, M. Ho, K. Lau, F. Cardona, and D. Hui, "Natural fibre-reinforced

composites for bioengineering and environmental engineering applications," *Compos. Part B Eng.*, vol. 40, no. 7, pp. 655–663, 2009.

[4] M. J. M. Nor, N. Jamaludin, and F. M. Tamiri, "Effect of perforated panel in acoustics absorption performance of coir fiber material," in *12th International Congress on Sound and Vibration 2005, ICSV 2005*, 2005.

[5] M. J. M. Nor, N. Jamaludin, and F. M. Tamiri, "A preliminary study of sound absorption using multi-layer coconut coir fibers," *Electron. J. Tech. Acoust.*, vol. 3, pp. 1–8, 2004.

[6] S. Ersoy and H. Küçük, "Investigation of industrial tea-leaf-fibre waste material for its sound absorption properties," *Appl. Acoust.*, vol. 70, no. 1, pp. 215–220, 2009.

[7] L. P. Bastos, G. da S. V. de Melo, and N. S. Soeiro, "Panels manufactured from vegetable fibers: An alternative approach for controlling noises in indoor environments," *Adv. Acoust. Vib.*, vol. 2012, 2012.

[8] H. Xiang, D. Wang, H. Liua, N. Zhao, and J. Xu, "Investigation on sound absorption properties of kapok fibers," *Chinese J. Polym. Sci.*, vol. 31, no. 3, pp. 521–529, 2013.

[9] A. Putra, Y. Abdullah, H. Efendy, W. M. Farid, M. R. Ayob, and M. S. Py, "Utilizing sugarcane wasted fibers as a sustainable acoustic absorber," *Procedia Eng.*, vol. 53, pp. 632–638, 2013.

[10] S. Prabhakaran, V. Krishnaraj, and R. Zitoune, "Sound and vibration damping properties of flax fiber reinforced composites," *Procedia Eng.*, vol. 97, pp. 573–581, 2014.

[11] S. Taj, M. A. Munawar, and S. Khan, "Natural fiber-reinforced polymer composites," *Proceedings-Pakistan Acad. Sci.*, vol. 44, no. 2, p. 129, 2007.

[12] M. J. John and S. Thomas, "Biofibres and biocomposites," *Carbohydr. Polym.*, vol. 71, no. 3, pp. 343–364, 2008.

[13] Z. Hong, L. Bo, H. Guangsu, and H. Jia, "A novel composite sound absorber with recycled rubber particles," *J. Sound Vib.*, vol. 304, no. 1–2, pp. 400–406, 2007.

[14] E. Gliscinska, M. Michalak, I. Krucinska, J. Kazimierczak, A. Bloda, and D. Ciechanska, "Sound absorbing composites from nonwoven and cellulose submicrofibres," *J. Chem. Chem. Eng.*, vol. 7, no. 10, p. 942, 2013.

MODELLING OF THE COUPLED BEAM-PIEZOELECTRIC MATERIAL WITH HYSTERESIS NON-LINERITY EFFECT

Mohd Hafiz Abdul Satar and Ahmad Zhafran Ahmad Mazlan*

TheVibrationLab, School of Mechanical Engineering, Engineering Campus, Universiti Sains Malaysia, Penang

*Email: zhafran@usm.my

ABSTRACT

Hysteresis is one of the non-linearity characteristics of the piezoelectric material. This characteristic is important to be characterized since it can affect the performance of the piezoelectric material as sensor or actuator in many applications. In this study, the model of the coupled aluminium beam with single piezoelectric patch material is constructed to investigate the hysteresis effect of the piezoelectric material to the whole beam structure. A P-876 DuraAct^{TM} type piezoelectric patch material is used in modelling of the piezoelectric actuator. Firstly, the modal analysis of the coupled beam-piezoelectric actuator is determined to get the natural frequencies and mode shapes. Then, the piezoelectric patch material is investigated in terms of actuator by given a sinusoidal voltage excitation and output in terms of deflection, stress and strain of the piezoelectric actuator are investigated. From the results, it is clear that, the coupled beam-piezoelectric material is affected by the hysteresis of the piezoelectric material and the natural frequencies of the beam structure. This characteristic is important for the piezoelectric actuator manufacturer and by providing the correction algorithm, it can improve the performance of the piezoelectric actuator for many applications.

Keywords: Hysteresis, coupled beam-piezoelectric material, modal analysis, voltage excitation, deflection.

INTRODUCTION

Hysteresis is one of non-linear characteristic of the piezoelectric. It has caused 'lag' between mechanical and electric relationship. The hysteresis loop itself contain more or less the information of magnitude response and phase angle produced by a periodic signal with constant amplitude [1]. For the application of piezoelectric, it has restrained the piezoelectric to act as an actuator, sensor or both.

From previous study, hysteresis is observed as a rate-dependent inherent characteristic and more likely to occur at high operating frequencies [2]. In term of error contribution to piezoelectric performance, it caused up to 15 % of tracking error to the total displacement range and will give disadvantage for any application [3].

In recent studies, modeling and dynamic response of the piezoelectric has been widely using finite element method (FEM). This method includes modal analysis, harmonic response and transient analysis that allow further understanding between mechanical systems and vibration effects [4].

In [5], the FEM method also has been used in characterization of double piezoelectric material as actuator. The piezoelectric is significantly capable to demonstrate as an actuator and output acceleration is increased when the voltage increased. This article also shows a good indicator whereby the output acceleration is maximum when close to the natural frequency.

METHODOLOGY

Modal Analysis of the Coupled Beam-Piezoelectric Material

Modal analysis of the coupled beam-piezoelectric is performed to determine the vibration characteristic such as natural frequencies and mode shapes [6]. The modelling of the beam structure is constructed as shown in Figure 1. The material properties of the beam are selected from the engineering data library and assigned with aluminum alloy as shown in Table 1. The ceramic type piezoelectric material used in this structural analysis is PIC255 and more importantly, the piezoelectric properties are directional-independent, or the pooling direction is consistent based on element coordinate system and material definition. In this analysis, the pooling direction is set in z-direction and data properties is summarized in Table 2.

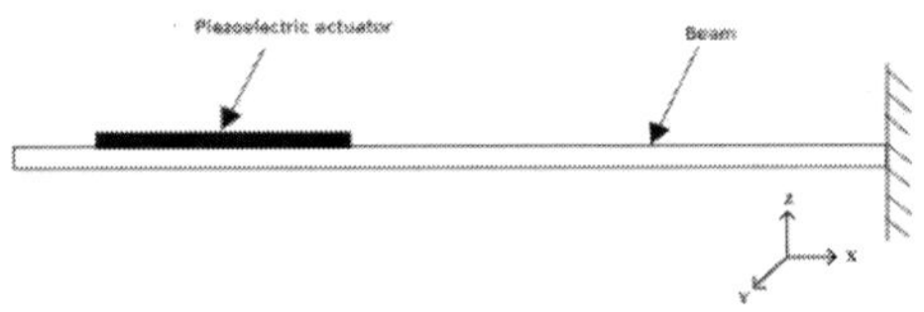

Figure 1: Model of coupled beam-piezoelectric [5].

Table 1: Aluminum alloy properties [5].

Property	Value
Density	2770 kg/m^3
Isotropic elasticity:	
1. Young's modulus, E	71000 MPa
2. Poisson's ratio, v	0.33
3. Bulk modulus, E_b	69600 MPa
4. Shear modulus, E_s	26692 MPa

Table 2: Piezoelectric patch properties.

Property	Value
Density, ρ	7800 kg/m3
Anisotropic elasticity, C	$\begin{bmatrix} 12300 & 7025 & 7670 & 0 & 0 & 0 \\ 0 & 9711 & 7025 & 0 & 0 & 0 \\ 0 & 0 & 12300 & 0 & 0 & 0 \\ 0 & 0 & 0 & 2226 & 0 & 0 \\ 0 & 0 & 0 & 0 & 2226 & 0 \\ 0 & 0 & 0 & 0 & 0 & 2315 \end{bmatrix} \times 10^{10}\ Pa$
Piezoelectric constant, e	$\begin{bmatrix} 0 & -715 & 0 \\ 0 & 1370 & 0 \\ 0 & -715 & 0 \\ 1190 & 0 & 0 \\ 0 & 0 & 1190 \\ 0 & 0 & 0 \end{bmatrix} N/Vm$
Permittivity constant, $\frac{\varepsilon}{\varepsilon_0}$	$\begin{bmatrix} 1649 & 0 & 0 \\ 0 & 1750 & 0 \\ 0 & 0 & 1649 \end{bmatrix}$

Piezoelectric Patch Material Characterization

In the analysis setting, the time step used is 0.00025 s with step end time of 0.1 s. For the piezoelectric actuator patch, the piezoelectric body is defined as MEMs body. The voltage input function is calculated from the equation below:

$$V = V_o \sin(2\pi f t) \quad (1)$$

Where $V_0 = 100 \sim 500\text{V}$ and $f = 100 \sim 500Hz$. For the results, a probe will be used and located at end of the beam. This will show the result (i.e. deformation, velocity and acceleration) varies over time and hysteresis data is collected from range of 0.02 s to 0.035 s. In addition, the solution result of the acceleration is set in *z*-direction due to beam structure is displaced in this direction.

RESULTS AND DISCUSSION

Modal Analysis and Harmonic Response of the Coupled Beam-Piezoelectric Material

The results of natural frequencies and mode shapes of the coupled beam-piezoelectric are shown in Figure 2. The natural frequencies are ranges from 0 to 500 Hz and four mode shapes obtained as shown in Figure 2. The natural frequencies are 16.61 Hz, 157.3 Hz and 546.4 Hz as first bending, second and third bending respectively. Whereas, the third mode shape is at 179.2 Hz with first torsional. The frequency response function (FRF) graph is obtained in harmonic response analysis. Figure 3 shows the FRF of acceleration at *z*-axis direction. Its shows that the bending modes at first and second modes are significantly affected the beam compared to torsion (third mode). Nevertheless, the pattern shows the acceleration tend to decrease as distance from frequencies at 16.61 Hz and 179.2 Hz.

First Mode (16.61 Hz)

Second Mode (157.3 Hz)

Third Mode (179.2 Hz)

Fourth Mode (546.4 Hz)

Figure 2: Mode shapes of beam structure.

 https://doi.org/10.1051/matecconf/201821702001

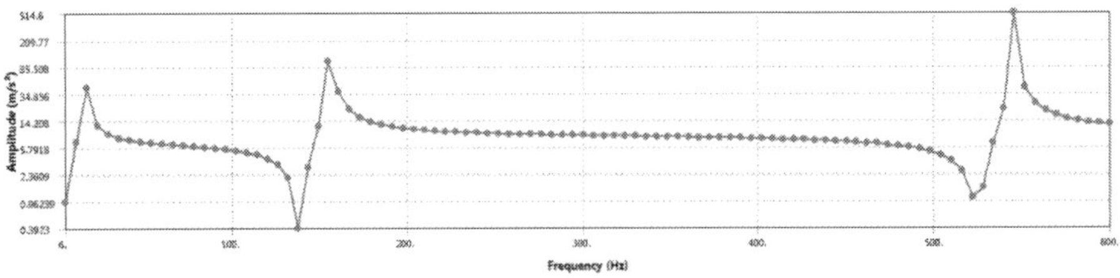

Figure 3: FRF of acceleration of the beam structure.

Piezoelectric Patch Material Characteristics with Hysteresis Effects

Figure 4 shows the illustration of the displacement, stress and strain analyses occurred in coupled beam-piezoelectric structure. In one complete cycle analysis, there are always a peak and a through point and each of this point has different stress and strain contours. At peak point in Figure 5(a) and Figure 6(a), the highest stress occurred around the piezoelectric mounted to the beam with 9.77 MPa and strain tend to elongate at fixed end of beam with 4.49×10^{-6}. For through point in Figure 5(b), it has small stress contour with only of 0.243 MPa. But differently for strain, it occurred in middle region of beam structure and small strain contour of 2.44×10^{-5}, as shown in Figure 6(b).

(a) (b)

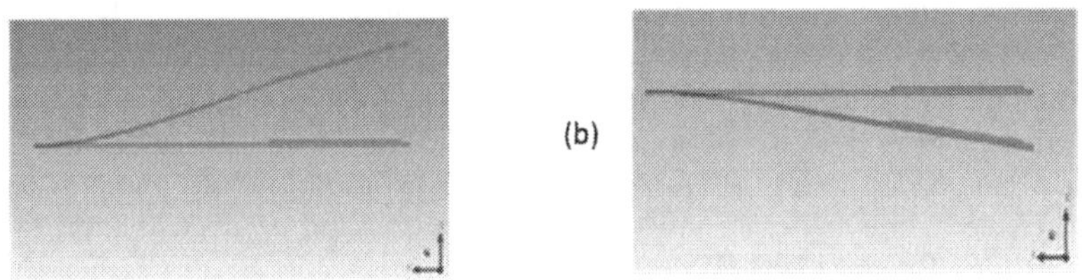

Figure 4: Deflection of beam structure at (a) peak point (b) through point.

(a) (b)

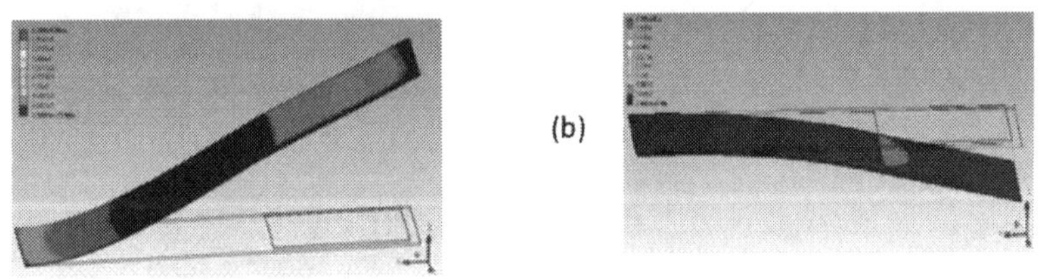

Figure 5: Stress contour at (a) peak point (b) through point.

(a) (b)

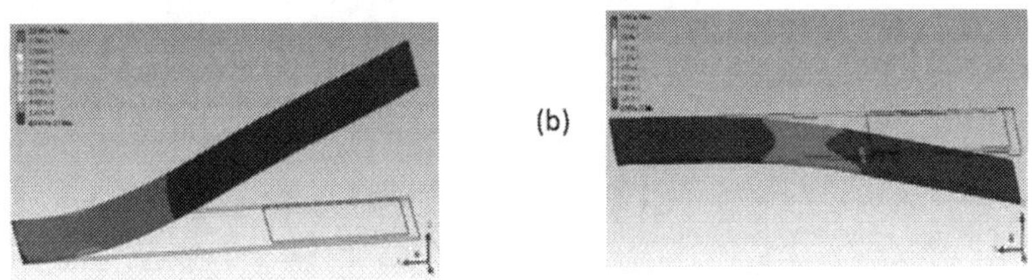

Figure 6: Strain contour at (a) peak point (b) through point.

For the transient analysis, the results obtained from ANSYS is tabulated and plotted as in Figure 7, the graphs of output acceleration against input voltage for one complete cycle of the piezoelectric patch actuator at different operation frequencies and voltages. In Figure 7(a), it is found that the hysteresis area at frequency of 100 Hz is getting larger when the input voltage increased. To be noticed, the hysteresis area is increased consistently except at negative voltage input voltage for 500 V.

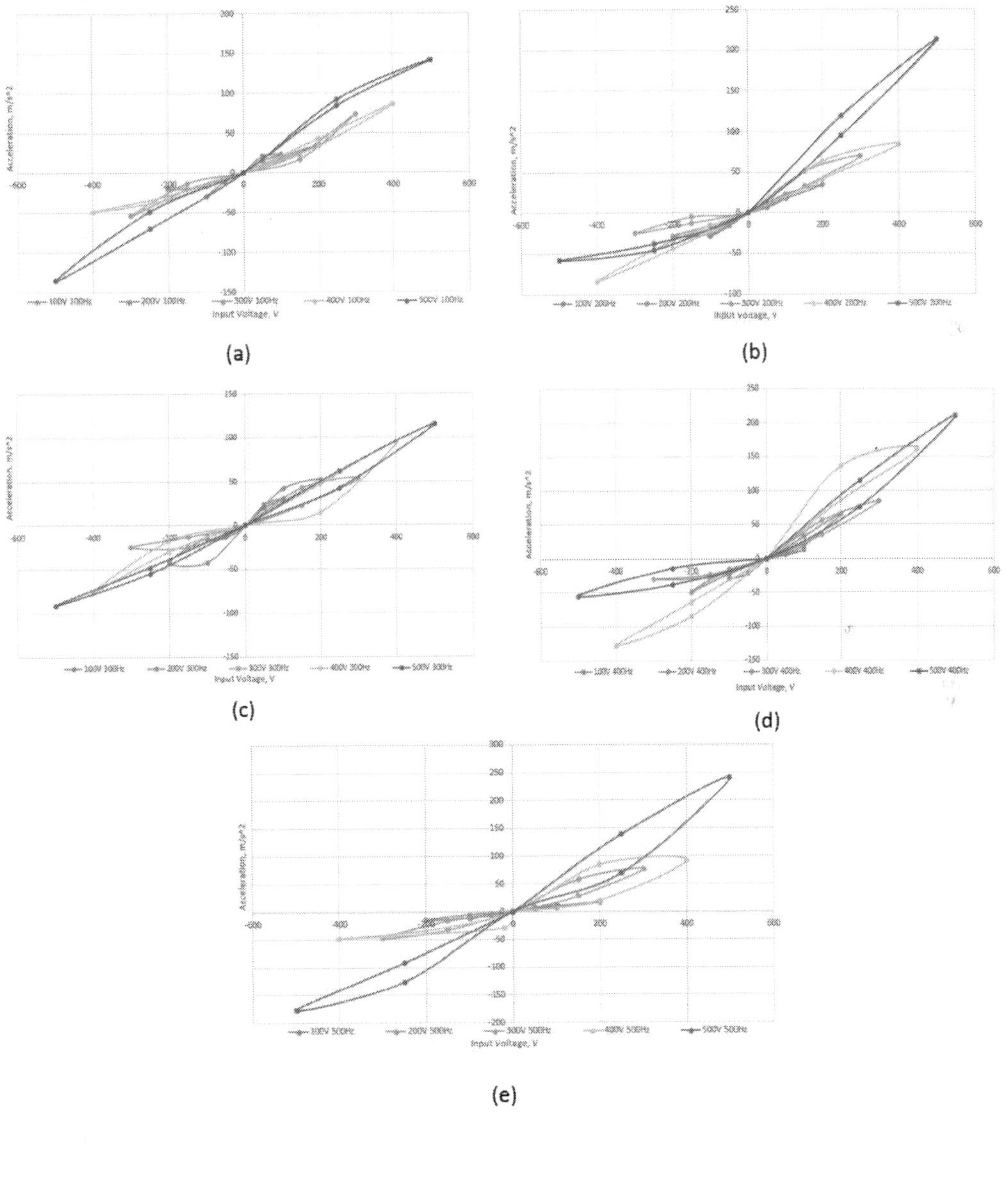

(a) (b)

(c) (d)

(e)

Figure 7: Hysteresis effect of the piezoelectric patch actuator for frequency of (a) 100 Hz (b) 200 Hz (c) 300 Hz (d) 400 Hz and (e) 500 Hz.

Similarly, for 200 Hz, the area of hysteresis at frequency of 200 Hz enlarges constantly as shown in Figure 7(b). The maximum acceleration produced at 200 Hz is 212 ms^{-2}, which higher compared to 100 Hz and 300 Hz. This acceleration produced may explained by the second natural frequency of beam structure by referring to Figure 3, previously. At 300 Hz in Figure 7(c), the area of hysteresis is larger especially at positive input voltages. However, hysteresis curve started to see obvious when operate at higher than 200 V.

For 400 Hz and 500 Hz in Figure 7(d) and Figure 7(e) respectively, the area of hysteresis is larger in positive input voltage which is similar as in Figure 7(c). In comparison, the hysteresis effect is expected to grow larger at 500 Hz compared to 400 Hz. Both graphs are expectedly to be expanding with increasing of input voltages.

CONCLUSION

Throughout this study, the objectives are achieved to generate the dynamic responses of the coupled beam-piezoelectric structure using FEM model. The deflection, stress and strain analyses show significant discoveries in terms of maximum and minimum contours. The modal and transient analyses show that, the hysteresis affected the performance of piezoelectric. Thus, the correction algorithms or methods should be applied to improve the output produced by the piezoelectric actuator.

ACKNOWLEDGEMENTS

The authors would like to acknowledge Universiti Sains Malaysia and Kementerian Pengajian Tinggi for providing the financial assistance under Short-term and FRGS grants (304/PMEKANIK/60313052 and 203/PMEKANIK/6071370).

REFERENCES

[1] D. Damjanovic, Hysteresis in piezoelectric and ferroelectric materials, vol. 3. 2006.

[2] Y. Liu, J. Shan, U. Gabbert, and N. Qi, "Hysteresis and creep modeling and compensation for a piezoelectric actuator using a fractional-order Maxwell resistive capacitor approach," Smart Mater. Struct., vol. 22, no. 11, 2013.

[3] K. K. Leang and S. Devasia, "Design of hysteresis-compensating iterative learning control for piezo-positioners: Application to atomic force microscopes," Mechatronics, vol. 16, no. 3–4, pp. 141–158, 2006.

[4] G. Takács, Active Vibration Control Prototyping in Ansys: A Verification Experiment, vol. 19. 2011.

[5] M. S. Vijaya, "Piezoelectric Materials and Devices: Applications in Engineering and Medical Sciences," vol. 186 pages, no. August, 2012.

[6] M. Imran, A. A. Abbasi, and M. J. Hyder, "Determination of modal characteristics of cantilever beam," ICET 2016 - 2016 Int. Conf. Emerg. Technol., vol. 2, no. 2, pp. 4–6, 2017.

STRUCTURAL DYNAMIC ANALYSIS OF THE CHENDEROH DAM SECTOR GATE SECTION

Mohamad Hazwan Mohd Ghazali[1], Mohd Hafiz Zawawi[2], Nurul Husna Hassan[2], Mohd Rashid Mohd Radzi[2,3], Ahmad Zhafran Ahmad Mazlan[1]*, Mohamad Aizat Abas[1] and Mohd Remy Rozainy Mohd Arif Zainol[4]

[1]School of Mechanical Engineering, Engineering Campus, Universiti Sains Malaysia, Penang
[2]Department of Civil Engineering, College of Engineering, Universiti Tenaga Nasional, Selangor
[3]TNB Generation Division, Tenaga Nasional Berhad, Kuala Lumpur
[4]School of Civil Engineering, Engineering Campus, Universiti Sains Malaysia, Penang
*Email: zhafran@usm.my

ABSTRACT

The dynamic characteristics such as natural frequencies, mode shapes and frequency response function (FRF) are the important characteristics to be investigated to access the level of durability of any dam structures. These characteristics are important since it will be the reference information for any operational methods to be used for the dam structures. In this study, one of the real dam (i.e., Chenderoh Dam) that available in Malaysia is taken into consideration, where the dynamic analysis of the sector gate section of the dam structure is investigated. The real scale of the sector gate section is measured on site and modelled into the CAD software with the consideration of real build-in materials. Then, the finite element (FE) model is constructed in ANSYS software with the required boundary condition and meshing sensitivity analysis. From the result of modal analysis, 30 natural frequencies are determined in the range of 0.5904 Hz to 8.471 Hz together with the mode shapes but only the most significant natural frequencies will be shown in this paper. In addition, all three axes of the FRF graphs show an agreement for the highest natural frequency value at 7.95 Hz, where the maximum deflection occurred in x axis direction with 2.03×10^{-7} m.

Keywords: Dynamic analysis, natural frequency, mode shape, FRF, dam structure.

INTRODUCTION

Sector gate section of the dam controls the flow of water from upstream to the downstream area and considered as one of the vital parts of the dam structure. Thus, it is important to study the dynamic characteristic of the dam in order to determine the structure behavior under dynamic loading. Usually when analyzing the dam, static and dynamic properties are considered as fundamental elements where static properties are related to the design and post-construction monitoring and dynamic properties take preference in structural monitoring of dam in seismic zones [1]. Natural frequencies, mode shapes and FRF are among the critical dynamic characteristics of the dam structure. Natural frequencies can be defined as frequencies at which the structure naturally inclines to vibrate if it is subjected to disturbance whereas the behavior of the structure at a specific natural frequency of vibration are known as mode shape [2]. In addition, each mode shape is corresponding to a specific natural frequency. FRF on the other hand helps to determine the response of a structure at a given frequency in terms of deformation, velocity and acceleration [3]. The peak responses of FRF graph are related to the natural frequencies of the structure [4]. Better understanding on dynamic response and information regarding the dynamic characteristic of dam structure will be useful in maintaining the safety of the dam structure [5].

METHODOLOGY

Initial Validation of Dam Material

Experimental modal analysis (EMA) is carried out for the real dam concrete material and compared with the simulation of two different concrete materials in order to validate the natural frequencies of the materials before running the simulation for the sector gate section. LMS TestLab, impact hammer and accelerometers are used for the EMA whereas simulation is performed using ANSYS software. A 3 cm x 3 cm x 3 cm cube of dam concrete material is extracted as a sample for the initial validation process.

Modelling of the Sector Gate Section of the Dam Structure

Sector gate section is located at the center of the overall dam. In sector gate section, the release of water at the spillway location is controlled by two gates, which is sector gate and radial gate. The three-dimensional (3D) model of the sector gate section is constructed using SolidWorks software. Figure 1 shows the real and 3D model of the sector gate section which consists of several important parts.

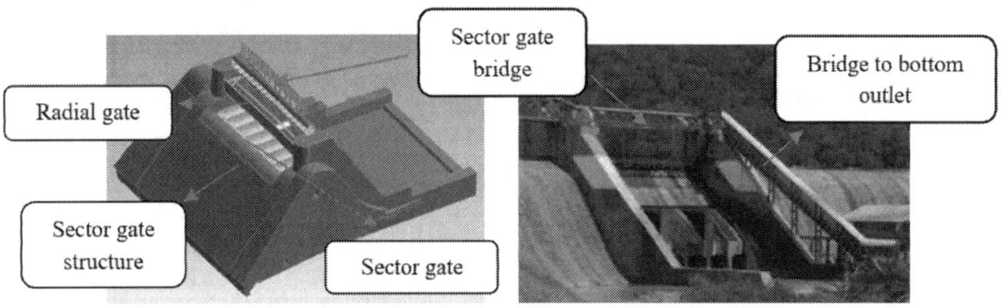

Figure 1: 3D model of sector gate section (left) and real picture of sector gate (right)

FE Model and Sensitivity Analysis

FE model of the sector gate section is developed using ANSYS software and it is made up of 433,404 tetrahedral elements. Sensitivity analysis is performed to obtain the optimized meshing type for the simulation. Three types of meshing which are hex dominant, tetrahedral and automatic mesh is investigated with the same body sizing. The boundary condition is set as fixed support for all the bottom faces. Figure 2 shows the FE model of the sector gate as well as its boundary condition.

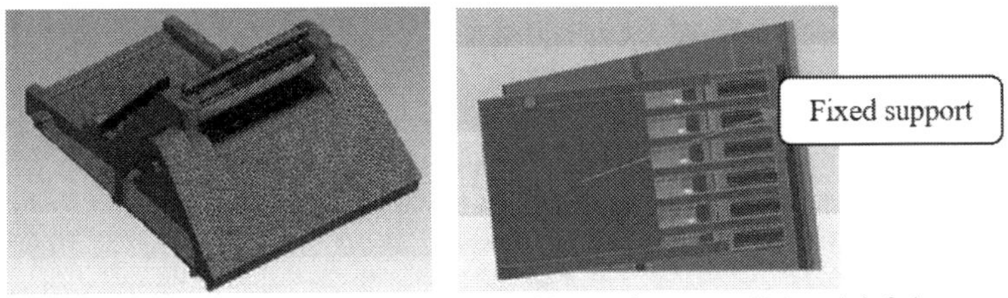

Figure 2: FE model of sector gate (left) and boundary condition (right)

Modal Analysis of the Sector Gate Section

Modal analysis of the sector gate section is performed in ANSYS software by importing the CAD file from SolidWorks software. Then, the required materials which are mild steel and concrete are added and assigned to respective parts in the sector gate section. Before starting the meshing process, all parts must share the same topology. 30 mode shapes are set as the maximum mode shapes to be determined in the this analysis.

RESULTS AND DISCUSSION

Initial Validation of Dam Material

The first natural frequency from the EMA of real dam concrete material is 6,570 Hz whereas simulation result gives the value of 6,400 Hz as the first natural frequency. By comparison, the percentage error between EMA and simulation is 2.59 %, which is valid for further analysis of the dam structure.

Sensitivity Analysis of the Sector Gate Section

The result of the sensitivity analysis shows that the hex dominant and tetrahedral mesh give the same first natural frequency which is 4.65 Hz compared to the automatic mesh which is at 4.72 Hz for the case of full concrete materials. Figure 3 shows the relationship between the natural frequency and types of mesh and it can be concluded that the hex dominant mesh is the optimized meshing type for the simulation.

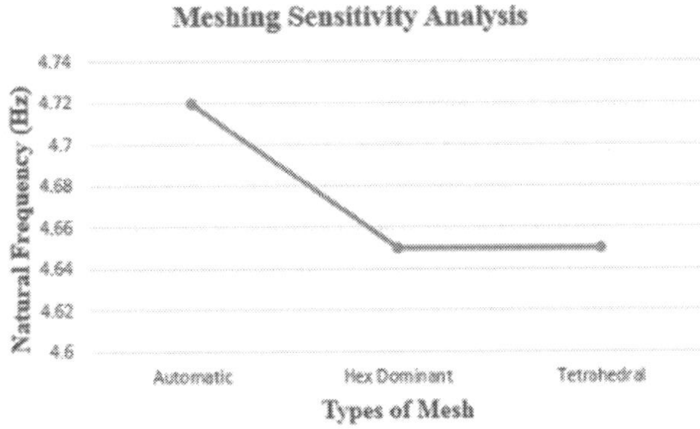

Figure 3: Sensitivity analysis of three different types of mesh (Full concrete)

Natural Frequency, Mode Shape and FRF of the Sector Gate Section

Table 1 shows 4 out of 30 mode shapes obtained along with its corresponding natural frequencies and deflection values. It can be observed that the 16th mode shape gives the most significant effect to the sector gate with the deflection value of 0.694 mm and natural frequency of 5.090 Hz. Any vibration near to this frequency will cause deflection to the sector and radial gate, sector gate bridge, upper section of the sector gate structure and the highest deflection will occur at the bridge to the bottom outlet. From the FRF graph of harmonic analysis, the highest natural frequency of the *x*, *y* and *z* directions occurred at 7.95 Hz with a maximum deflection of 2.03 x 10^{-7} m, 2 x 10^{-7} m and 1.79 x 10^{-7} m, respectively. The highest deflection occurred at *x* axis direction for the sector gate section. Table 2 shows all three FRF graph in *x*, *y* and *z* directions.

Table 1: First and three most significant mode shapes with its respective natural frequencies and deflection values

Modes No.	Mode Shapes	Natural Frequency	Deflection Values
1		0.5904 Hz	28.41 mm
7		1.908 Hz	2.79 mm
16		5.090 Hz	0.694 mm

https://doi.org/10.1051/matecconf/201821702002

30		8.471 Hz	1.75 mm

Table 2: FRF graphs for the sector gate section in *x*, *y* and *z* directions

Direction	**FRF graph**
x	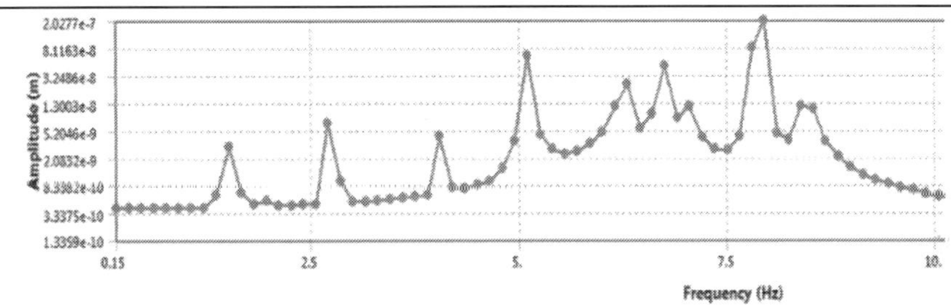
y	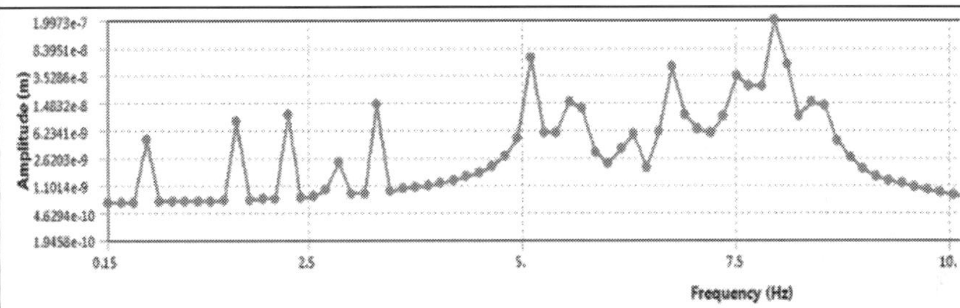
z	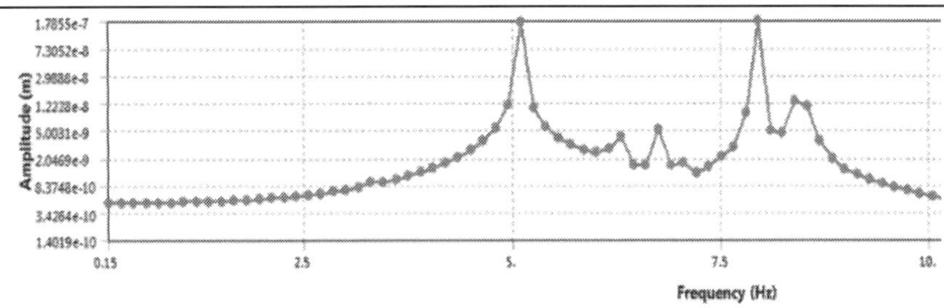

CONCLUSION

From this study, structural dynamic analysis of the Chenderoh Dam Sector Gate Section has been successfully carried out. It can be concluded that:

- The development of the 3D model of the sector gate section has been successfully done.

- The meshing sensitivity analysis has been conducted to determine the optimum and reliable meshing for the sector gate and the hex dominant meshing is selected as the most optimum meshing.

- The modal analysis result of the sector gate section which include the natural frequencies and mode shapes have been successfully achieved with the 16th mode shape at 5.09 Hz gives the most significant effect to the overall sector gate section.

- The harmonic response analysis result of the sector gate section with the FRF graph has been successfully achieved with the maximum deflection occur in *x* direction at 7.95 Hz.

ACKNOWLEDGEMENTS

The authors would like to acknowledge Universiti Sains Malaysia and Uniten R&D Sdn. Bhd. for providing the facilities and financial assistance under accounts (304/PMEKANIK/60313052 and U-841).

REFERENCES

[1] Vezi M. (2014). Dynamic Modelling of Arch Dams in the Ambient State. *Doctoral Dissertation*, University of Cape Town.

[2] Chapter 3: Real Eigenvalue Analysis (2005). Retrieved from: *http://www2.me.rochester.edu/courses/ME443/NASTRAN/Chpt3RealEigenvalue Analysis.pdf.*

[3] Chapter 10: Harmonic Analysis. (2005). *ANSYS Training Manual.* Inventory #00221510-4.

[4] C. M. Ramesha, Abhijith K. G., Abhinav Singh, Abhishek Raj and Chetan S Naik. (2015). Modal Analysis and Harmonic Response Analysis of a Crankshaft. *International Journal of Emerging Technology and Advanced Engineering,* Volume 5, Issue 6.

[5] Shanshan W. and Qingwen Ren. (2011). Dynamic Response of Gravity Dam Model with Crack and Damage Detection. *Science China Technological Sciences*, *54*(3), 541-546.

INVESTIGATION OF DAMPING COEFFICIENT FOR MAGNETORHEOLOGICAL ELASTOMER

M.N.Hazwan Hadzir[1], Z.A.Norfaidayu[1], M.S.Mohamad Sabri[1], Muhamad-Husaini Abu-Bakar[1*]

[1]System Engineering and Energy Laboratory,
Universiti Kuala Lumpur,
Malaysian Spanish Institute,
Kulim Hi-tech Park,09000,
Kulim,Kedah, Malaysia.
*Email: muhamadhusaini@unikl.edu.my
Phone: +6044035199; Fax: +6044035201

ABSTRACT

Magnetorheological elastomer (MRE) is a smart material where its stiffness vary with presence of magnetic field.These special properties lead to various application as mechanical devise such as damper. Damping coefficient is criteria that need to consider while designing the device. This paper aims to study the effect of iron filling and elastomer composition on a damping coefficient of MRE. Five MRE specimens with different CIP composition from 30 to 70% was prepared.the speciment was tested with testbed which is has a Constance distance permanent magnet(5mm) and motor speed (131.9Hz). The vibration amplitude in frequency domain was used to calculate the damping coefficient by half-power bandwidth equation. Experimental result indicates that the highest damping coefficient is at 30% CIP concentration and the lowest is at 70% CIP composition. As a conclusion, the damping coeficiet is decrease with increasing in CIP composition.this finding useful to design an optimal active damper for mechanical component.

Keywords: Magnetorheological elastomer, smart material, damping coefficient, semi-active damper, damping properties.

INTRODUCTION

MRE was categorized as an intelligent material due to ability in reclaims its shape when a magnetic field is removed. MRE physical state can be tuned between a soft elastomer and a semi-solid with varying an external magnetic field induced to the material(Chen & Liao, 2007). MRE has overcome some the problem on MRF, such as deposition, sealing issue and environmental contamination. The pioneer work on the magnetically sensitive elastomer was firstly reported by (Rigbi & Jilkén, 1983).

MRE consist of three necessary components: magnetic particle (carbonyl iron), an elastomer, and additives (usually silicon oil.). The magnetic particle size such as carbonyl iron commonly is lie in between 5-10 micron. The Elastomer such as silicon rubber used as they are soft or deformable at room temperature, the elastomer can have the ability to extend from 5-700%, depending on the specific material. (Ruddy, Ahearne, & Byrne, 2012). The component is mixed together to form a compound with significant density magnetic particle randomly dispersed or pre-arrange in a low-density matrix. In the absence of magnetic field, the magnetic particle will attract each other as shown in figure 1.

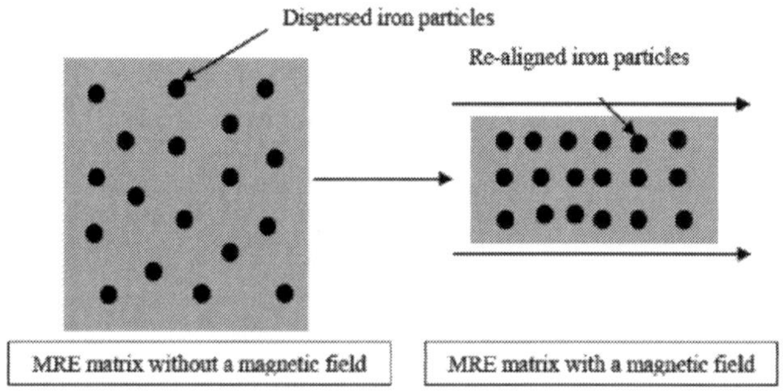

Figure 1. Schematic diagram of MRE (Usman & Jung, 2015)

Magneto-rheological elastomer(MRE) has their specialties compare with MR fluid and conventional device nowadays (Wereley, 2013). MRE can work in multiple degrees of freedom (DOF) while MR fluid work in a single degree of freedom. Magnetorheological elastomer device work in the pre-yield region, while MR fluid work in the post-yield region. Thus, MRE usually can work with variable stiffness while MR Fluid device works in variable damping. Due to able work in variable stiffness when the magnetic field applied, the most researcher looking potential in MRE to develop smart devices such as vibration isolation, a vibration absorber, magnetic sensor and actuator (Y. Li, Li, Li, & Du, 2014).

The mechanical and rheological properties of an elastomeric material containing magnetizable particles can be changed reversibly by applying an external magnetic field. This is called the magnetorheological effect. The origin of the field dependence of the MRE properties is the existence of field-induced dipole magnetic forces between the particles. Therefore, the mechanical behavior of MR elastomer is a combination of matrix properties and the properties of magnetizable particle formations inside the matrix.

Dynamic mechanical properties testing has been used for characterizing the shear storage modulus of MR elastomer (Ismail, Ibrahim, Hamid, Mahmood, & Adnan, 2016). Most of the test have been done by the use of force oscillation (Huang, Yan, Liu, Li, & Wang, 2015). Recently, free vibration techniques have also been used in MRE studies (G. H. Li, Huang, Gu, & Wang, 2013). Perhaps, the researcher investigated the modulus storage of the MRE and shear stress of the material to characterize the properties of the material. G. H. Li et al., (2013) was propose the new method of preparation MRE specimen and new approach to identify the mechanical properties of MRE. Shuib & Pickering, (2016) use Dynamic mechanical properties testing to identify dynamic properties of MRE based on iron sand and natural rubber. Kukla, Górecki, Malujda, Talaśka, & Tarkowski, (2017) use compression method to identify the behavior and properties of MRE. MRE was considered as a smart material because it is able to control the stiffness by controlling the magnetic field. Thus, the investigation of damping coefficient for MRE is needed for optimizing the application in the featured.

This paper, isotropic MRE specimen was prepared by various percentage of Carbonyl Iron Particle (CIP) composition from 30% to 70%. The accelerometer sensor was used to measure the vibration in every CIP composition. The damping coefficient was calculated by half-power bandwidth equation.

METHODOLOGY

Magnetorheological Elastomer (MRE) Preparation

In this section, the preparation of Magnetorheological Elastomer specimens for experimental are explained. The specimen was prepared by following the portion in table 1; there are four material components were used in the preparation of MRE specimens, which is Carbonyl iron particle (CIP), Natural silicon rubber, grease and silicone oil. The portion of composition was allocated following the volume of MRE mold in this experiment.

Table 1. The composition of MRE specimens

% of CIP	Rubber(gram)	Grease(gram)	Silicone Oil (gram)	CIP(gram)
30	84.9	6.06	12.13	36.4
40	88.5	7.38	14.76	59.02
50	86.92	8.69	17.32	86.92
60	80.04	10.00	20.01	120.06
70	67.91	11.32	22.63	158.46

The MRE specimen fabrication procedure starts with mixing the grease and the natural silicone rubber in a bowl during 10min to obtain an elastomer gel with good homogenization. Next, pour silicon oil and continues mixing in 10 minutes until the mixer homogenize. Lastly, pour mixing for 15 minutes until it is homogenization. The mixer MRE is loaded into the mold and stored for 24 hours at room temperature.

Experimental Setup

Figure 2 shows the experimental setup for investigation of MRE damping coefficient. Figure 2(a) shows the controller component for controlling the motor speed including the Arduino, MD10C motor driver and battery as an energy source for Arduino. Figure 2(b) is the MRE damping coefficient setup; this setup has three component, DC motor as vibration source, MRE specimens, and a permanent magnet. The piezoelectric accelerometer sensor was used to measure the vibration. The signal from accelerometer sensors are recorded using LMS SCADAS hardware, and then the data is analyzed using LMS Test.Lab software to generate the vibration figures and graphs data. The experiment was conducted by several of MRE specimens from 30% to 70% CIP filling.

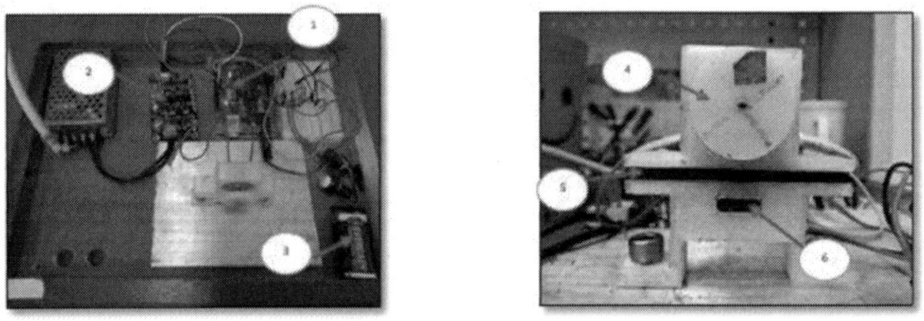

(a) (b)

Figure 2. (a) Controller setup(1)Arduino,(2)MD10C Motor Driver,(3)Battery; (b) vibration setup (4)DC motor,(5)MRE specimen,(6)Permanent Magnet

Half-power bandwidth equation.

The equation to calculate the damping coefficient is by using the half-power bandwidth method in equation 1.

$$\frac{\omega_2 - \omega_1}{\omega_n} = 2\zeta \quad (1)$$

Where ω(omega) is angular frequency, and ζ(zeta) is damping coefficient.

RESULTS AND DISCUSSION

Frequency Domain Analyses

Figure 3 shows the vibration amplitude comparison by various CIP composition. The data was collected by LMS SCADAS hardware and piezoelectric accelerometer sensor. The data was analyzed by LMS Test.lab to generate frequency domain graph for further investigation. The graph in figure 3 shows, all the components are difference amplitude for every specimen. The higher amplitude is at 70% CIP composition with pick value 0.116g. The lowers amplitude in this graph is at 30% CIP composition with pick value 0.0147g. The amplitude keeps increasing proportional increasing the CIP composition due to the increase of stiffness. However, the non-linearity was detected in 60% CIP composition; the amplitude drastically drop lower than 50% CIP composition and 40% CIP composition event the test was repeated in 5 times, his phenomena is the new contribution in MRE for further study.

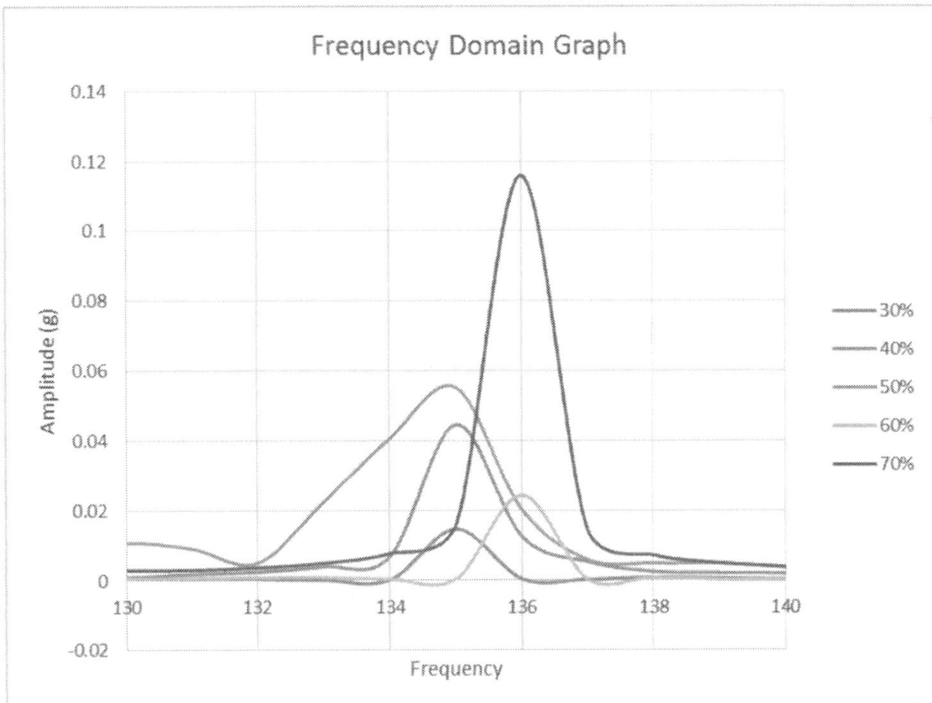

Figure 3. Vibration comparison

Figure 4 shows the damping coefficient value for a different CIP composition in MRE. The damping coefficient value calculates as explained in equation 1. The graph in figure 4 shows that maximum value of damping coefficient occurs at CIP with 30% and the value was 0.1988. There was no significant difference in damping coefficient value when the CIP percentage varied from 40% to 60%. However, the damping coefficient value dropped drastically when the CIP in MRE was 70%.the span of damping coefficient was 0.0728 which is maximum 0.1988 and minimum 0.126. The factor that contributes to the high value of damping coefficient for CIP 30% is it stiffness are lowest as compare to other CIP composition. When the CIP % is high, the magnetic field permeability is increased, and the internal force in the MRE increase. The force is increased because the distance between the CIP is closer and create a better magnetic flux flow path.

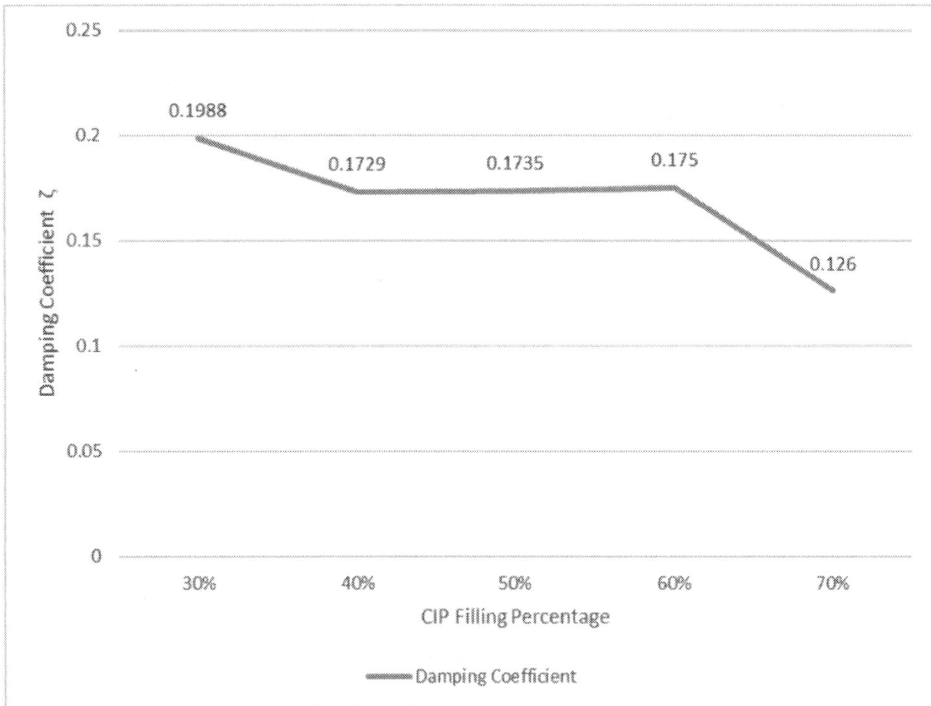

Figure 4. Damping coefficient

CONCLUSION

Experiment result reveals the effect of CIP on damping coefficient of MRE. MRE with 30% CIP shows the highest damping coefficient value compare to other composition. The stiffness was decreased with increases in CIP composition in MRE. As a conclusion; the MRE damping coefficient was inversely proportional to CIP composition. This finding useful to design a mechanical device such as damper and vibration isolator. Because the by magnetic field influence the MRE damping coefficient, the experiment should extend to the different level of magnetic field magnitude.

ACKNOWLEDGEMENTS

The authors would like to acknowledge Universiti Kuala Lumpur, Malaysian Spanish Institute, System Engineering and Energy Laboratory UniKL Msi and Dynamic Vibration Laboratory UniKL Msi for providing the consultation, technical support, and give permission to use the equipment facilities in their Laboratory for make sure this project is successful.

REFERENCES

[1] Chen, J., & Liao, W. H. (2007). Design and control of a magnetorheological actuator for leg exoskeleton. *2007 IEEE International Conference on Robotics and Biomimetics, ROBIO*, 1388–1393. https://doi.org/10.1109/ROBIO.2007.4522367

[2] Huang, X. G., Yan, Z. Y., Liu, C., Li, G. H., & Wang, J. (2015). Study on the resistance properties of magnetorheological elastomer. *Materials Research Innovations*, *19*(sup5), S5-924-S5-928. https://doi.org/10.1179/1432891714Z.0000000001223

[3] Ismail, R., Ibrahim, A., Hamid, H. A., Mahmood, M. R., & Adnan, A. (2016). Dynamic mechanical behavior magnetorheological nanocomposites containing CNTs: A review. *AIP Conference Proceedings*, *1733*. https://doi.org/10.1063/1.4948878

[4] Kukla, M., Górecki, J., Malujda, I., Talaśka, K., & Tarkowski, P. (2017). The Determination of Mechanical Properties of Magnetorheological Elastomers (MREs). *Procedia Engineering*, *177*, 324–330. https://doi.org/10.1016/j.proeng.2017.02.233

[5] Li, G. H., Huang, X. G., Gu, X. Y., & Wang, J. (2013). Fabrication and Mechanical Properties Study of the Magnetorheological Elastomer. *Applied Mechanics and Materials*, *376*, 148–152. https://doi.org/10.4028/www.scientific.net/AMM.376.148

[6] Li, Y., Li, J., Li, W., & Du, H. (2014). A state-of-the-art review on magnetorheological elastomer devices. *Smart Materials and Structures*, *23*(12), 123001. https://doi.org/10.1088/0964-1726/23/12/123001

[7] Rigbi, Z., & Jilkén, L. (1983). The response of an elastomer filled with soft ferrite to mechanical and magnetic influences. *Journal of Magnetism and Magnetic Materials*, *37*(3), 267–276. https://doi.org/10.1016/0304-8853(83)90055-0

[8] Ruddy, C., Ahearne, E., & Byrne, G. (2012). A review of Magnetorheological Elastomers: Properties and Applications. *Advanced Manufacturing Science (AMS) Research*, *20*, 1–7.

[9] Shuib, R. K., & Pickering, K. L. (2016). Effect of Carbon Black on the Dynamic Properties of Anisotropic Magnetorheological Elastomer, *12*, 1–12.

[10] Usman, M., & Jung, H.-J. (2015). *Recent developments of magneto-rheological elastomers for civil engineering applications*.

[11]Wereley, N. M. (2013). *Magnetorheology*. https://doi.org/10.1039/9781849737548

DEVELOPMENT OF A LOWER LIMB STROKE REHABILITATION MACHINE

Khuan Foo Law[1], Dalbir Singh[1] and Zaidi Mohd Ripin[1]

[1]TheVibrationLab,
School of Mechanical Engineering,
Universiti Sains Malaysia,
14300 Nibong Tebal, SPS,
Pulau Pinang, Malaysia.
*Email: mezaidi@usm.my
Phone: +6045996368; Fax: +6045996912

ABSTRACT

This paper explains the development of a lower limb stroke rehabilitation machine designed for subacute stroke patients. The system is capable of measuring the muscle force and providing goal-oriented feedback in real-time and running in two different rehabilitation modes. The mean value of engagement metric of healthy subjects using this machine with feedback was 24.53% higher than without feedback. This proved that feedback can help the patients to be fully engaged during the rehab session and this can be useful in strengthening the neuromotor pathways. The brain recovery based on the motor cortex correlation quantification algorithm based on the electroencephalography (EEG) signals which is validated against the established technique based on the functional magnetic resonance imaging (fMRI). From the results, the resting-state EEG beta coherence of healthy subjects was found to be 0.474±0.06, whereas the average fMRI functional connectivity between left and right primary motor areas of healthy subjects was 0.537±0.08. The percentage difference was only 11.7%. clinical trial will be carried out to further measure the efficacy of the rehabilitation treatment using this system.

Keywords: electroencephalography, functional connectivity, functional magnetic resonance imaging, lower limb, motor cortex, recovery, rehabilitation, stroke, subacute and visual feedback.

INTRODUCTION

According to the World Heart Federation, 15 million people worldwide suffer from stroke and nearly five million are left permanently disabled with paralysis and weaknesses every year [1]. For instance, approximately 50,000 Canadians and 780,000 Americans suffer a stroke each year, One-sided paralysis is a common effect of stroke which diminishes the strength and control of the lower limb [2,3]. About 80% of patients with stroke experience motor weakness/hemiparesis [4].

The functional recovery and long-term health status of the stroke patients are more affected by rehabilitation during the subacute stage [5]. Generally, the recovery from stroke occurs within 3 months after the onset of stroke. To be more specific, it is proven that the neurological and functional recovery of stroke patients occurs at the higher rate during subacute stage [6]. Rehabilitation at the sub-acute stage is difficult due to the fact that the patients are generally not strong enough to carry out exercise of the lower limb. In this project, a small, lightweight and portable rehabilitation machine is developed specifically for the stroke patient to carry out rehabilitation exercise which still on the bed. Commonly, stroke rehabilitation guidelines recommend the intensive and repetitive practice of functional tasks after stroke [7]. Moving the limb in a manner that self-generated effort cannot achieve, assistive exercise provides novel somatosensory stimulation that helps induce brain plasticity [8]. Studies [9-10] suggest that providing too much assistance may have negative consequences for learning. Example strategies to encourage participant effort and self-initiated movements by implementing the active resistive exercise, triggering assistance only when participants achieve a force or velocity threshold, making robot compliant [11]. The brain is able to rebuild the damaged neuron paths through the conscious control of the limb's motion [12] and this can be achieved through movement re-education [13, 14, 15]. In terms of quantification of stroke recovery, it is commonly indicated by Fugl Meyer Assessment (FMA) score. It is a stroke-specific, performance-based impairment index and is designed to assess motor functioning, coordination, balance, sensation and joint functioning in patients with post-stroke hemiplegia [16]. However, its scoring is based on direct observation of performance and subjective evaluation and broad based. Hence, a more specific and direct measurement is needed and in this project, the push force of the lower limb during flexion-extension movement can be obtained.

In order to tackle all the stated problems related to the rehabilitation of the lower limbs for stroke patients, three main objectives have been highlighted as below:

1. To design a lower limb rehabilitation machine which can monitor muscle strength recovery.
2. To investigate the engagement of stroke patients in rehabilitation exercises.

METHODOLOGY

Design of the lower limb rehabilitation machine

The view of the machine with all the major components is as shown in Fig. 1 below. In this figure, the machine is shown with load cells which are used to measure the force applied by the lower limb of stroke patients throughout the rehabilitation exercises, as an indication of the lower limb motor performance. Meanwhile, ultrasonic sensors are used for the measurement of the range of motion, and eventually to provide real-time visual feedback on the screen.

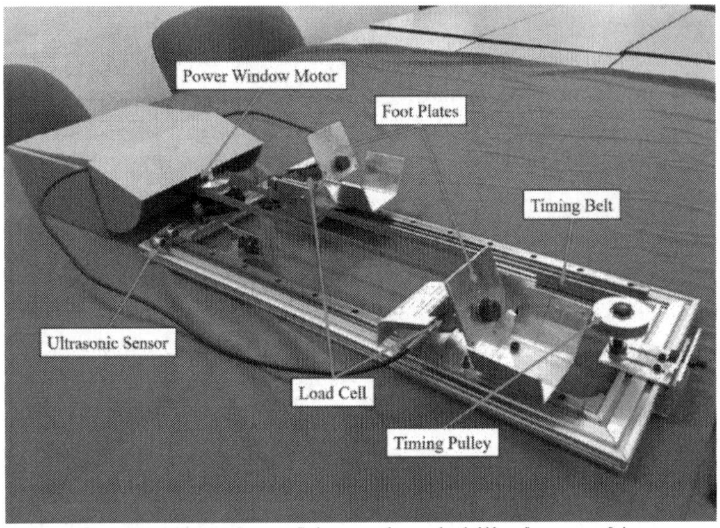

Figure 1. Overview of the stroke rehabilitation machine

The intelligence of this machine can be defined from its capability to measure muscle force, provide real-time visual feedback as well as the ability to run in two different modes of exercises. The two different modes in which the machine can run, are namely Continuous Passive Motion (CPM) and Active Resistive Motion (ARM). In CPM mode, the physiotherapist just has to select the CPM mode and key in the number of cycles required. The system is provided with an emergency stop button to protect the patient in case of emergency. In the ARM mode, the force from the foot of the motion measured using the load cells will be compared to a range of set values, within which the machine will keep on moving. If the push force from the feet is lower or higher the system will stop and advise the patient to lower or increase the force from the foot.

RESULTS

Fig.2 below shows the iLLSRM system prototype being tested with healthy patient. In this setup the patient is requred to push the device and at the same time paying attention to the display on the screen which requires the patient to exert the force in some predetermined level. Fig. 3 shows the real time push force from the healthy leg and the paretic leg which allows for direct comparison of the difference of the push force. This will allow for better communication and also monitoring of patient recovery.

 https://doi.org/10.1051/matecconf/201821702004

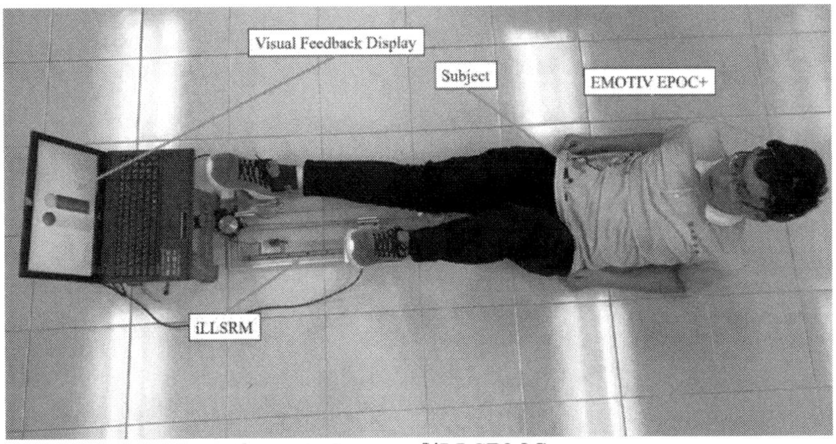

Figure 2. Set-up of iLLSRM System

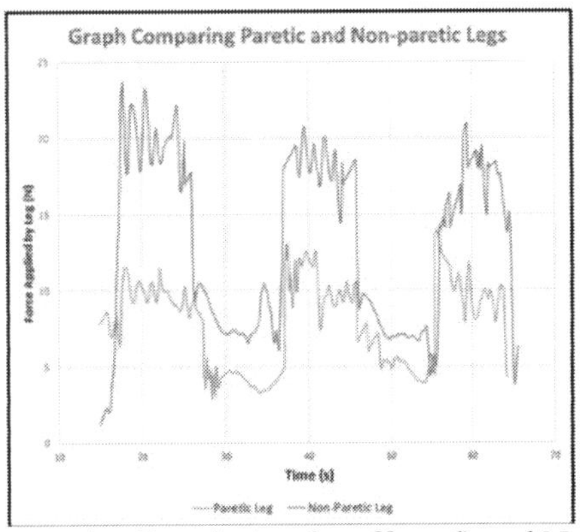

Figure 3. Display Comparing Paretic (stroke affected) and Non-Paretic Leg (non-affected)

Measuring Engagement

The results of the engagement measure of all the five healthy subjects with age 21–24 (mean = 22.8 ± 1.3 years), undergoing exercises with and without visual feedback is shown in the box plot in Fig. 4 below to compare the engagement measure before and after thefeedback was introduced in the exercise.

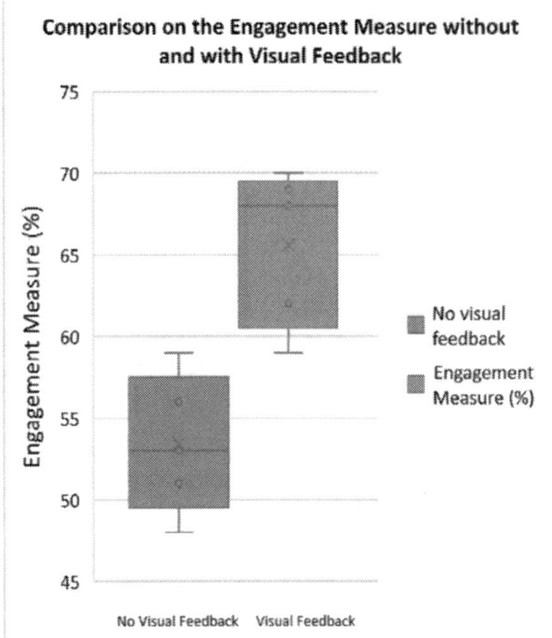

Figure 4. Engagement Measure Comparison for Exercises

DISCUSSION

The machine has shown its intelligence and capability in monitoring muscle strength recovery, displaying real-time visual feedback and running in two different modes. As shown in Fig. 3, the force exerted by the non-paretic limb is used as a baseline strength to indicate the full recovery of the paretic limb. This is because the strength of the lower limb varies from one person to another. The paretic limb is considered to recover fully once it is able to exert the same magnitude of force as the non-paretic limb. When it is being observed carefully, there is a slight lag off in the graph on force exerted by the paretic limb.

The percentage of increment in engagement measure was 24.53%. Hence, it can be deduced that patients will be more concentrated and highly engaged in their rehabilitation exercises as compared to conventional machines. This is due to the fact that the system interactively communicates to the patients concerning the muscle performance in terms of their range of motion, as well as the minimum force which they have to overcome for the continuous movement. It can maintain the engagement of patients in undergoing the therapy without or with lesser feeling of boredom, as there are certain targets to be achieved in their rehabilitations. At the same time, it will be stimulating the somatosensory cortex in the affected hemisphere, which is known to be involved in cortical reorganization after stroke.

CONCLUSION

This project has successfully come out with a lower limb stroke rehabilitation machine which can provide goal-oriented visual feedback in real-time, measuring lower limb muscle force and running in either Continuous Passive Motion (CPM) or Active Resistive Motion (ARM). It is also proven that visual feedback can help patients in being more engaged in their rehabilitation.

ACKNOWLEDGEMENTS

The authors wish to thank the five healthy subjects, namely D. M. Lau, Y. X. Tan, J. D'raj, K. F. Law and D. Singh, who were involved in the acquisition of resting-state EEG data as well as the experiment of investigating the effect of real-time visual feedback on the engagement. They are all currently the students of Engineering Campus, Universiti Sains Malaysia.

REFERENCES

[1] Berry, S.M.M.R.V.T., B.R.A.P. Wilson, and L.F.J.M.D.P.M.P.H. Fallon, *Strokes*. 2017, Salem Press.

[2] Lindsay, P., et al., *Toward a more effective approach to stroke: Canadian Best Practice Recommendations for Stroke Care*. Canadian Medical Association Journal, 2008. 178(11): p. 1418-1425.

[3] Daniel, K., et al., *What Are the Social Consequences of Stroke for Working-Aged Adults?* Stroke, 2009. 40(6): p. e431.

[4] Pollock, Alex & Baer, Gillian & Campbell, Pauline & Ling Choo, Pei & Forster, Anne & Morris, Jacqui & Pomeroy, Valerie & Langhorne, Peter. (2014). Physical rehabilitation approaches for the recovery of function and mobility following stroke. The Cochrane database of systematic reviews. 4. CD001920. 10.1002/14651858.CD001920.pub3.

[5] Jung HY. (2017) Rehabilitation in Subacute and Chronic Stage After Stroke. In: Lee SH. (eds) Stroke Revisited: Diagnosis and Treatment of Ischemic Stroke. Stroke Revisited. Springer, Singapore

[6] Rønning, O.M., & Guldvog, B. (1998). Outcome of subacute stroke rehabilitation: a randomized controlled trial. *Stroke, 29 4*, 779-84.

[7] Hebert, D., Lindsay, M. P., McIntyre, A., Kirton, A., Rumney, P. G., Bagg, S., et al. (2016). Canadian Stroke Best Practice Recommendations: Stroke Rehabilitation Practice Guidelines, Update 2015. International Journal of Stroke, 11(4), 459-484. doi: 10.1177/1747493016643553

[8] Poon, C.-S. Sensorimotor learning and information processing by Bayesian internal models. in Engineering in Medicine and Biology Society, 2004. IEMBS'04. 26th Annual International Conference of the IEEE. 2004. IEEE.

[9] Israel, J.F., et al., Metabolic costs and muscle activity patterns during robotic-and therapist-assisted treadmill walking in individuals with incomplete spinal cord injury. Physical therapy, 2006. **86**(11): p. 1466-1478.

[10] Wolbrecht, E.T., et al. Real-time computer modeling of weakness following stroke optimizes robotic assistance for movement therapy. in Neural Engineering, 2007. CNE'07. 3rd International IEEE/EMBS Conference on. 2007. IEEE.

[11] Marchal-Crespo, L. and D.J. Reinkensmeyer, Review of control strategies for robotic movement training after neurologic injury. Journal of neuroengineering and rehabilitation, 2009. **6**(1): p. 20.
[12] Kwakkel, G., B.J. Kollen, and H.I. Krebs, Effects of Robot-Assisted Therapy on Upper Limb Recovery After Stroke: A Systematic Review. Neurorehabilitation and Neural Repair, 2007. **22**(2): p. 111-121.
[13] Thikey, Heather Anne and van Wijck, Frederike and Grealy, Madeleine and Rowe, Philip (2011) A need for meaningful visual feedback of lower extremity function after stroke. In: 2011 5th International Conference on Pervasive Computing Technologies for Healthcare (PervasiveHealth 2011). Institute of Electrical and Electronics Engineers, pp. 379-383. ISBN 9781612847672
[14] Li, Chong & Rusák, Zoltán & Horvath, Imre & Ji, Linhong. (2016). Validation of the reasoning of an entry-level cyber-physical stroke rehabilitation system equipped with engagement enhancing capabilities. Engineering Applications of Artificial Intelligence. 56. 185-199. 10.1016/j.engappai.2016.09.006.
[15] N. Lehrer and L. Olson (2009). Visual feedback for Mixed Reality stroke rehabilitation, *2009 Virtual Rehabilitation International Conference*, Haifa, pp. 194-194.
[16] Physiopedia. *Fugl-Meyer Assessment of Motor Recovery after Stroke.* Retrieved fromhttps://www.physio-pedia.com/Fugl
Meyer_Assessment_of_Motor_Recovery_after_Stroke

AN EXPERIMENTAL STUDY ON THE EFFECT OF SURFACE TEXTURE ON CONTACT STIFFNESS USING SOFT MATERIAL

Z. Fuadi

Center for Applied Acoustic, Thermal, and Vibration Studies,
LPPM Syiah Kuala University
Gedung Fakultas Teknik Universitas Syiah Kuala Lt. 1
Jl. Tgk. Tanoh Abee Darussalam Banda Aceh 23111 INDONESIA
*Email: zahrul.fuadi@unsyiah.ac.id

ABSTRACT

Contact interface is one of the most important factors in a mechanical contact because it is the place where friction, sound, and heat originate. It is therefore inevitable that modeling various phenomenon related to contact dynamics requires a proper representation of the contact interfaces. One of the methods in representing the behavior of two surfaces in contact is by using the parameter of contact stiffness. In this study, the effect of surface texture on contact stiffness is analyzed. The texture was used in order to reduce the randomness of surface roughness. The soft material was chosen to achieve a pure elastic contact thus preventing plastic deformation to the asperities. The analysis was conducted by using an indentation method employing a steel ball with a relatively small indentation force. The result show contact stiffness values of the textured surfaces were smaller than that of smooth surface. This is particularly observed at low normal load at which total deformation of the surface is relatively small compared to the asperities height. This decrease in the contact stiffness value of the textured surfaces can be related to the reduction in the real contact area.

Keywords: contact stiffness, surface texture, soft material.

INTRODUCTION

The behavior of a mechanical contact interface is highly complex because it involves complex interaction among surface asperities influenced by various factors such as asperity size, shape, elastic properties, applied load, and the presence of lubricant at the interfaces. One of the parameters that incorporates most of the factors in a quantitative representation of a contact interface is the contact stiffness. The parameter has been incorporated in various analysis of dynamic problems such as brake noises [1,2,3,4,5] and other dynamic instability problems.

Among the factors affecting contact stiffness is the surface roughness, a parameter decided by the asperity size, spacing, and pattern. In a rough surface, the asperity size, spacing, and pattern are highly random so that there is no single parameter that can be used to quantify a surface roughness perfectly. Parameters such as arithmetic average of roughness profile Ra, Root Mean Square (RMS) of surface measured peaks and valleys, and the standard deviation of asperity height σ are among the parameters that are commonly used to represent a surface properties. The latter parameter has been incorporate in the theoretical estimation of the contact stiffness value of a contact interface [4]

Even though the contact stiffness parameter has been used in assessing various dynamic problems, the measurement of contact stiffness value of a contact interface is still difficult. Various methods have been proposed to experimentally measure the contact stiffness [6,7,8]. The real contact area, which is represented by the numbers of asperities in contact, is proportionally related to the contact stiffness value [6].

In this paper, the effect of surface texture on the quantitative value of normal contact stiffness is discussed. The contact stiffness value was assessed experimentally by using an indentation of a steel ball on a textured soft material surface. The effect of asperity height and spacing on the contact stiffness is highlighted.

METHODOLOGY

The contact stiffness, K_c, resulting from the interaction of the two planes is defined as the load, W, necessary to produce a unit displacement, d, of the mean separation and can be written as:

$$K_c = \frac{\delta W}{\delta d} \tag{1}$$

The load W is defined by:

$$W = \frac{1}{2\sqrt{\pi}} E(\sigma / \beta)^{1/2} A \tag{2}$$

in which E is the Young's modulus of the material, A is the contact area, β is the radius of curvature of asperity height, and σ is the standard deviation of asperity height. In a rough surface, the parameter of β and σ are highly random. The effect of parameter A and σ to the contact stiffness can be evaluated experimentally by using a surface with a relatively uniform contact asperities; surface texture. In order to find out the effect of those parameters on contact stiffness, various surfaces with different texture were prepared. In order to have a pure elastic contact, the soft material was used.

Surface Preparation

Four type of surfaces with different texture characteristics were prepared in this analysis. The profile of each surface is given in Figure 1. Each surface is marked as S-number i.e. S1, S2, S3, and S4. The surfaces were prepared by casting the silicon on copper surfaces with different textures, which were manufactured beforehand. Surface S1 (Figure 1a) was prepared by casting the silicon on a smooth copper surface. Surfaces S2, S3, and S4 were prepared by casting the silicon on the textured copper surfaces which were manufactured by scratching of a diamond pin with different spacing and depth. The surface S2 has an average asperity height of 15 μm and average asperity spacing of 0.07 mm. The surface S3 has an average asperity height of 20 μm and average asperity spacing of 0.14 mm while the surface S4 has an average asperity height of 40 μm and average asperity spacing of 0.25 mm.

Since the surfaces are made of similar material, they have a similar modulus of elasticity of 3 MPa. The surface will be made in contact with a steel ball having diameter of 2 mm. Thus, it is expected that the number of asperities in contact with the ball is different for each surface because of different asperity spacing. Thus, the change in contact stiffness values can be related to the number of asperity in contact and/or the real contact area.

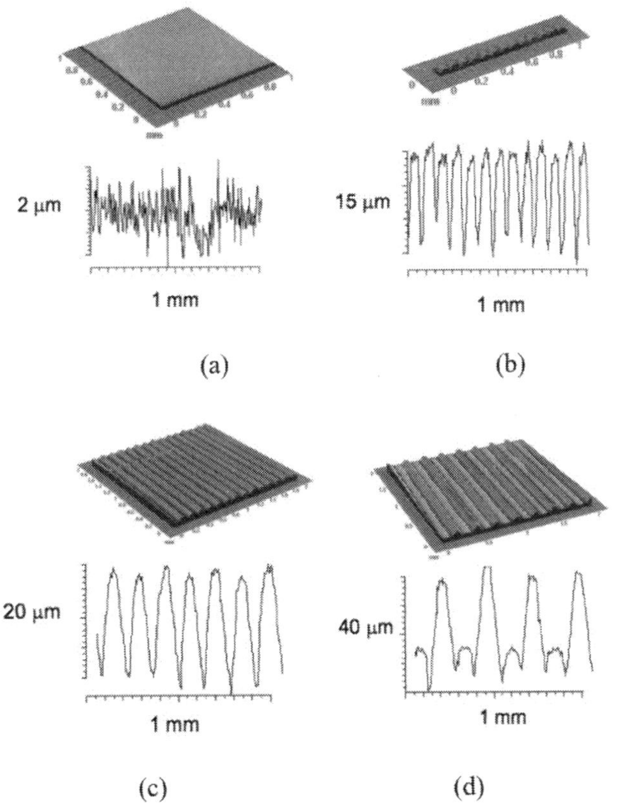

Figure 1. Profiles of surfaces used in this analysis; (a) surface S1, (b) surface S2, (c) surface S3, and (d) surface S4

Experimental Setup

The contact stiffness value was evaluated by using an indentation mechanism illustrated in Figure 2. The indenter ball has a diameter of 2 mm. A force is applied in normal direction so that the ball is in contact with the surface. A representative graph of an indentation is given in Figure 3. In the figure, a normal force of 12 mN was applied on the indenter resulting in the displacement of about 45 μm. Using the hysteresis data, the contact stiffness, Kn, is calculated as the ratio of force, ∂F, to displacement, ∂x, at the linear region as illustrated by Figure 3. Silicon material was used because it is highly elastic as well as to avoid plastic deformation.

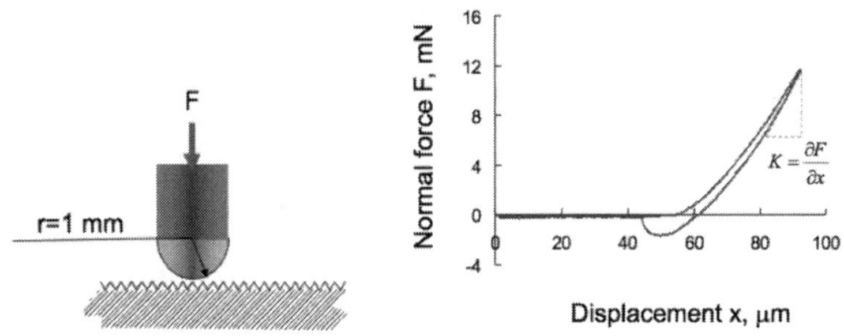

Figure 2. (a) Schematic illustration of the indentation and (b) an indentation graph

RESULTS AND DISCUSSION

Figure 3 shows the relationship among the normal force, mN, the Hertzian contact area, m^3, and the contact stiffness, N/m^2, for surface S1. It is shown in the figure that both the contact stiffness and the contact area increase as the effect of normal force increase. The indentation tests were conducted for the normal force of 30 mN, 60 mN, and 100 mN. At normal force of 30 mN, the contact area was $2.79\text{x}10^{-4}$ m^3, corresponding to the contact stiffness value of 859 N/m. The contact area increases to $4.61\text{x}10^{-4}$ m^3 at normal force of 100 mN resulting in the contact stiffness value of 1080 N/m. In this case, it seems that the contact area is linearly increases with the normal force but the contact stiffness is exponentially increases with the normal force increase.

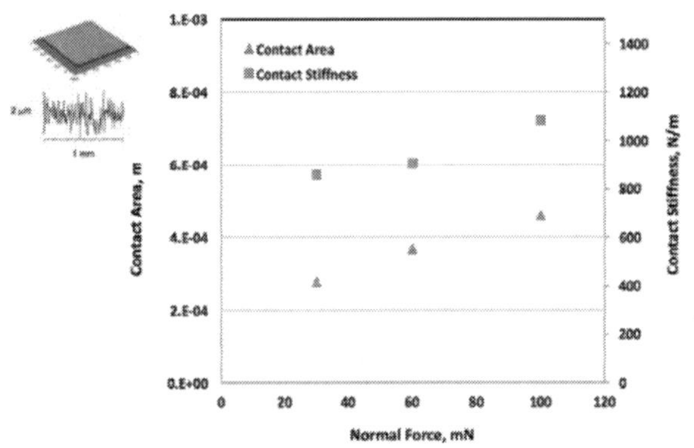

Figure 3. Relationship among the normal force, N, the Hertzian contact area, m^3, and contact stiffness, N/m of surface S1

Figure 4(a) shows the relationship among the normal force, mN, and the Hertzian contact area, m^3, of all evaluated surfaces. Since this value is calculated theoretically, there is no significant differences among the value of contact area of the evaluated surfaces. There is no difference between the contact area of the smooth surface and that of the textured surface for a similar normal force. For the normal force of 30 mN, the Hertzian contact area is about 0.0003 mm^3. The contact area increases to about 0.00046 mm^3 at normal force 100 mN.

However, the value of contact stiffness for each surface were different, as shown in Figure 4(b). At normal force of 30 mN, surface S1 has the highest value of contact stiffness, i.e. 859 N/m. The contact stiffness value of surfaces S2, S3, and S4 are 581 N/m, 493 N/m, and 586 N/m, respectively. This indicates that the contact stiffness is significantly affected by the surface texture

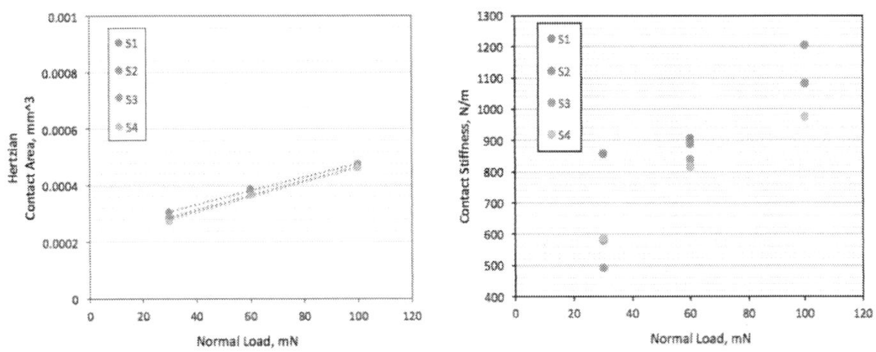

(a) (b)

Figure 4. (a) Relationship between the normal load, mN, and the Hertzian contact area, m^3, and (b) contact stiffness, N/m^2.

At normal force of 60 mN, the highest value of contact stiffness was for the surface S1 with a value of 906 N/m followed by that of surface S2 with a value of 890 N/m. The contact stiffness value for those of surface S3 and S4 are 839 N/m and 815 N/m, respectively. At the normal force of 60 N, the range of contact stiffness value for the tested surfaces is smaller than that at the 30 N. The highest value was achieved for the smooth surface thus indicating some effects of surface texture on the contact stiffness.

However, at the value of normal force 100 mN, the effect of surface texture on contact stiffness is less observable. At this point, the highest value of contact stiffness was achieved for surface S2 with a value of 1200 N/m. Surface S1 and S3 have a similar value of contact stiffness, i.e 1080 N/m while surface S4 has a contact stiffness value of 976 N/m. This is presumable due to high deformation during the indentation, as shown in Figure 5. At 100 mN the normal force caused a deformation more than 100 μm to the tested surfaces. This deformation is far larger than the height of the surface asperities as indicated in Figure 1. Nevertheless, the lowest value of contact stiffness was achieved for surface S4; the one that has the largest value of asperity height and spacing. This indicates the effects of asperity height and spacing parameters on the contact stiffness. In the future, it is interesting to analyze the quantitative relationship among these parameters on the contact stiffness.

The effect of surface texture on contact stiffness is summarized in Figure 6. The effect can be clearly observed at the normal force of 30 mN. At this force, the average deformation of the surface due to indentation force is 30 μm. With this value of deformation, it can be seen that surface S1 had the largest real contact area compared to other surfaces. Therefore, the decrease of contact stiffness value of surface S2, S3, and S3 can be related to the decrease of the real contact area, as shown in ref. [6].

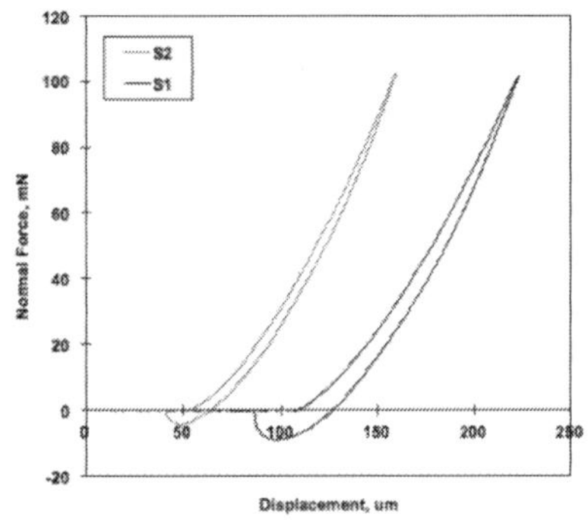

Figure 3 Representative results of the indentation tests

The effect of surface texture, asperity height and spacing, was less observable at normal force of 60 mN. At this normal force, the indentation caused deformation of more than 60 μm, which is larger than the highest value of asperity height of surface S4. The values contact stiffness were relatively similar for all surfaces, ranging from 815 N/m to 906 N/m. At this condition, it is presumable that the real contact area for all surfaces subjected to indentation were similar.

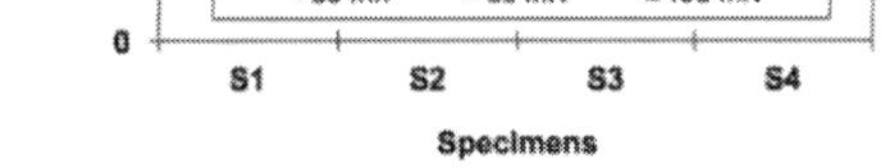

Figure 6 Effect of surface texture on the contact stiffness

A relatively similar condition was achieved for the indentation force of 100 mN, which resulted in surface deformation of more than 100 μm. In this condition, it seems there is another factor affecting the contact stiffness beside the real contact area. Further study is required to clarify this phenomenon.

It has been shown in previous studies that a contact interface having lower value of contact stiffness is less prone to stick-slip vibration. In this analysis, it is shown that a lower value of contact stiffness can be effectively achieved by reducing the real contact area, which is obtained by surface texturing.

CONCLUSION

An experimental analysis on the effect of surface texture on normal contact stiffness using soft material has been conducted. It is found that the values of contact stiffness for textured surface were smaller than that of smooth surface. It seems that the surface texture caused a reduction in the real contact area. This is particularly observed at low normal load at which total deformation of the surface is relatively small compared to the average asperities height. At higher value of normal load where surface deformation is larger than the asperity height, the effect of texture on normal contact stiffness was less significant. Further study is required to clarify the phenomenon,

REFERENCES

[1] K. Popp. Some model problems showing stick-slip motion and chaos, friction-induced vibration, chatter, squeal and chaos, (Ibrahim R. A. and Soom, A. ed), ASME, (1992) 1-12.
[2] W. W. Twordzydlo, E. E. Becker, J. T. Oden. Numerical modeling of friction-induced vibration and dynamic instabilities, friction-induced vibration, chatter, squeal and chaos, (Ibrahim R. A. and Soom, A. ed), ASME, (1992) 87-97.
[3] Z. Fuadi, K. Adachi, H. Ikeda, H. Naito, K. Kato. Effect of Contact Stiffness on Creep-Groan Occurrence on a Simple Caliper-Slider Experimental Model. Tribol Lett (2009) 33: 169
[4] Z. Fuadi, S. Maegawa, K. Nakano, K. Adachi. Map of low-frequency stick–slip of a creep groan. Proceedings of the Institution of Mechanical Engineers, Part J: Journal of Engineering Tribology Vol 224, Issue 12, pp. 1235 – 1246, 2010
[5] Z. Fuadi. Analysis of Vibration Generated by the Rubbing of Flat Surfaces. Makara Journal of Technology, 18, 3, pp. 115-120, 2015
[6] Z. Fuadi, T. Takagi, H. Miki, K. Adachi. An experimental method for tangential contact stiffness evaluation of contact interfaces with controlled contact asperities. Proceedings of the Institution of Mechanical Engineers, Part J: Journal of Engineering Tribology. Vol 227, Issue 10, pp. 1117 – 1128
[7] M. Eriten, C. H. Lee, A. A. Polycarpou, Measurements of tangential stiffness and damping of mechanical joints: Direct versus indirect contact resonance methods, Tribology International, Volume 50, 2012, Pages 35-44,
[8] S. Medina, D. Nowell, D. Dini. Analytical and Numerical Models for Tangential Stiffness of Rough Elastic Contacts. Tribology Letters. 2013;49:103-15.
[9] Z. Fuadi, H. Zahouani, T. Takagi, H. Miki. IOP Conference Series: Materials Science and Engineering 352 (1) pp. 012009

MODELING AND EXPERIMENT OF ELECTROMAGNETIC ENERGY HARVESTER SYSTEM BY USING DUAL MOVING MECHANICAL SYSTEMS AT LOW FREQUENCY RANGE.

M. Z. A. Rahim, M. N. H. Hamid*, Z. N. M. Yusuf, S. N. M. Soid and M. R. Ibrahim

Universiti Kuala Lumpur Malaysian Spanish Institute,
Kulim Hi-Tech Park
*Corresponding: mnajib@unikl.edu.my

ABSTRACT

This paper presented a concept of single degree of freedom (SDOF) electromagnetic vibration energy harvester device. This technique enable system to operate at wideband frequency range, low frequency and has multi-resonance frequencies. Each mechanical system operates at difference frequency where each system is attached with electromagnetic transducer components. The device is developed based on the parameter factors of vibration transmissibility from external vibration sources into the device through mathematical modelling. A prototype was tested by using vibration exciter and normal human walking. The fabricated device had showed multi-resonant behavior at 4.26 and 8.34 Hz during test. From experiment results, they have showed capability to operate at wide bandwidth frequency from 1.9 until 18.5 Hz at a periodic excitation of 0.04 g. The highest amount of rms voltage that has been produced about 108 mV with a maximum 78 μW average power across the 150 Ω load resistance. So, it has proven the dual-moving mechanical concept with low damping value in system has increased the operating bandwidth frequency and also increased the amount of output voltage from device.

Keywords: Parameter factor, Vibration transmissibility ratio, Electromagnetic vibration energy harvester system.

INTRODUCTION

Scavenging energy from mechanical vibration become the most study for harvesting research area due to its high availability in technical environments[1]. Vibrations have garnered the most interest because of their abundance, high theatrically achievable conversion efficiency and vibration transmissibility through different media [2]. The main mechanisms in energy harvester system are mechanical system and type of transducer. Mechanical mechanism system is more on physical design which involved the motional design [3] and a transducer is device that converts mechanical motion into electrical energy [4]. The transducer type commonly used in vibration energy harvesting is electromagnetic[2][5]. The capability of system to operate at wideband frequency depends on the mechanical design of system itself. In order to increase the system efficiency, many researchers are focusing on developing system with capability to operate under wideband frequency condition and had caused the varieties in design of system has been introduced [6][7][8].

In this paper, we developed electromagnetic vibrational energy harvester to improve the performance of the linear harvester based on the parametric factor involved in mechanical design in order to operate the system at time-varying frequency.

MOTIVATION AND SYSTEM DESIGN

Challenges in mechanical system for energy harvester.

Bandwidth and linearity of energy harvester are common problem faced by many researchers. Narrow bandwidth implies constrained resonance frequency and limit applications. The range of frequency depends on the efficiency of the system linearity. The mechanical design of energy harvester system will influence the capability of the system to operate at bandwidth frequency range. The lack of focusing in designing of electromechanical system (EMS) caused the limitation to fully utilize vibration level from the external sources (ambient sources) where reduce capability of energy harvester system to amplify the vibration level that indirectly it will influence the performance of system. This problem related to mechanical system mechanism. The efficiency of energy harvester system is determined based on amount of output voltage and ability to operate in wide operating frequency range.

Architecture, mechanism and its operation of proposed system

Mechanical system in energy harvester system is depend on parameter factors such as block mass (m), spring constant (k), and damping coefficient (d). In this research, the mechanical system of harvester has been developed based on parameters factor and linear behavior. Figure 1 illustrated the design system which had fully assembled. It also consists of electromagnetic system: magnet (3) and copper coil (4). All components involved in design are marked with numerical symbol as depicted in Figure 1 and explanation for each of components detail has been stated in table 1.

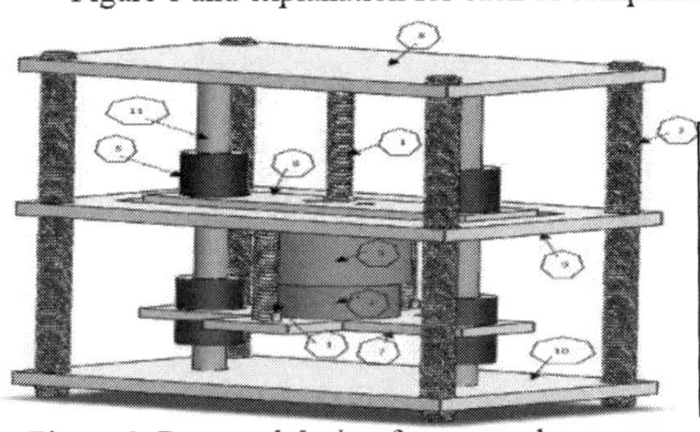

Figure 1. Proposed design for energy harvester system

Table 1. Components description of energy harvester system

No.	Components.	No.	Components.
1	Compressed spring	7	Lower stage
2	Thread road	8	Upper fixture
3	Magnetic core	9	Middle fixture
4	Copper coil	10	Base fixture
5	Linear bearing	11	Linear shaft
6	Upper stage		

In the system, mechanical system was divided into two mechanical subsystems which were upper system and lower system as depicted in Figure 2. Direction motion for mechanical subsystems has been fixed to operate in horizontal motion. Since the system used electromagnetic system as transducer, the motion between magnet and copper coil are very significant on determine the energy harvester system performance. That why, in proposed design has been built with two moving mechanical system in order to increase the performance of system. The upper system was attached with magnet core while the lower system was attached with copper coil as shown in Figure 2.

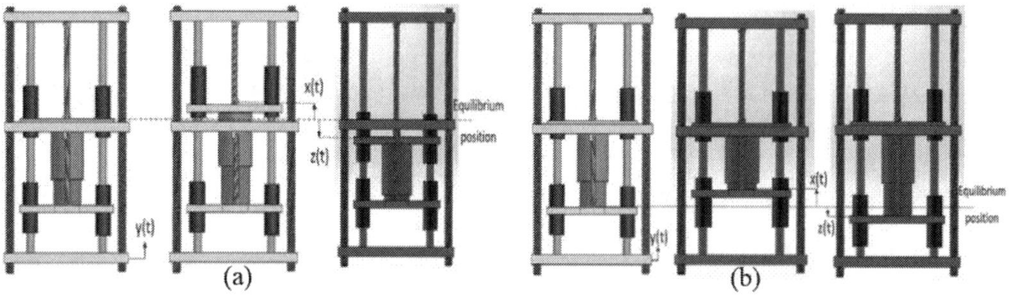

Figure 2. Illustration motion between magnet and copper coil for (a) upper stage and (b) lower stage.

The motion of upper x(t) and lower z(t) systems happen when external frequency y(t) is applied onto the system and since the electromagnetic components were attached together with mechanical subsystems, the motion of mechanical subsystems had caused the magnetic induction between magnet and coil happened which indirectly produced voltage as output. This technique can increase operating frequency range and output energy.

THEORETICAL MODELING AND PARAMETERS DESIGN.

Modeling of the mechanical design.

Mathematical modelling was developed were involving relationship between parametric factor on vibration transmissibility effect.

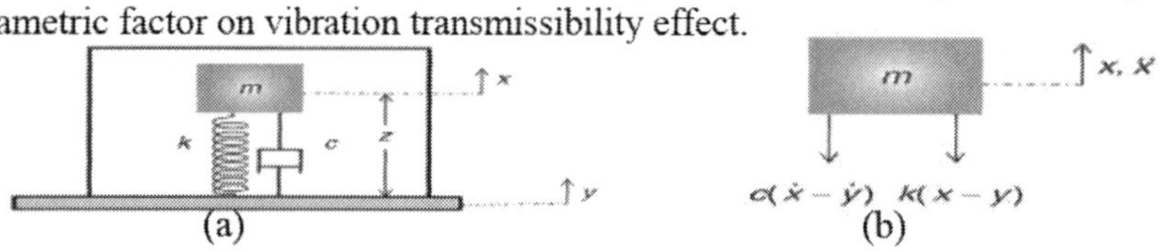

Figure 3: Schematic for SDOF of spring-mass-damper;(a) system was connected to external sources; (b) free-body diagram at an arbitrary instant including effect of support displacement, y (t).

Based in Figure 3, m is seismic mass (kg), k is spring stiffness (N/m), c is damping coefficient (N.s/m), y is displacement of external force, x is absolute displacement of seismic mass and z is relative deflection of the system. According to Newton's second law, the standard form of differential equation of mass relative deflection motion can be written as

$$\ddot{z} + 2\xi\omega_n\dot{z} + {\omega_n}^2 z = \omega^2 A \sin \omega t \quad (1)$$

Where ω is forcing frequency (rad/s), ξ is damping ratio and ω_n is angular frequency of the system. Since the oscillation of the response is at the same frequency as the excitation, hence the steady state relative response can be expressed as

$$z(t) = A_1 \sin \omega t + A_2 \cos \omega t \quad (2)$$

Differential Eq.2 into $\dot{z}(t)$ and $\ddot{z}(t)$,then substituted into Eq.1 and then simplified the solution into two equations where $r = \frac{\omega}{\omega_n}$, $z = \sqrt{A_1 + A_2}$ and the end of equation become

$$\frac{z}{y} = \frac{r^2}{\sqrt{(1-r^2)^2+(2\xi r)^2}} \quad ; \quad A_1 = \frac{F_0(k-m\omega^2)}{(k-m\omega^2)^2+c\omega^2} \quad ; \quad A_2 = \frac{F_0 c\omega^3}{-(k-m\omega^2)^2-c\omega^2}$$
(3)

The crucial of monitoring the relative deflection z of the system with respect to the base y is to control the reflective deflection of the system during operating. The application of vibration transmissibility in electromagnetic vibration energy harvester system was introduced by William and Yates [9].From the Eq.3, the effectiveness of vibration transmissibility for energy harvester system can be controlled with three type of parameters which are damping coefficient (c), seismic mass (m) and material stiffness(k).

Numerical simulation.

In figure 4 shows the numerical simulation result (MATLAB) for vibration displacement transmissibility based on the Eq. 3.

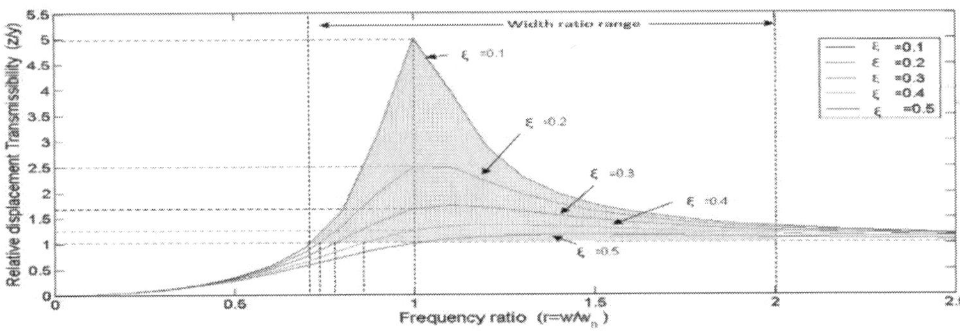

Figure 4. Relative displacement transmissibility with damping ratio (ξ) below 0.5.

Based on the graph in figure 4, smaller ξ value has increased width frequency ratio range and displacement transmissibility (z/y). It will give advantage for system to operate at any frequency ratio range without depend only on resonance condition (r=1). Thus, width range frequency ratio will help system to increase the performance of displacement transmissibility (z/y) even at varying external frequency value (ω). The frequency ratio (r= $\omega/\omega n$) below than value 1 and displacement transmissibility (z/y) less than value 1 are not suitable for system to operate in this region since it will limit the vibration transfer to system and indirectly it will give effect on system performance.

Prototype fabrication.

In order to proof the proposed idea, a harvester prototype was fabricated and tested. Figure 5 shows the image of fabricated prototype and the geometric parameter of components have been listed in Table 2.

Figure 5. Prototype of energy harvester system

Table 2. Geometric parameter of components.

Parameter	Value	Parameter	Value
Lower system mass	35.5 g	Upper system mass	67 g
Overall system mass	150 g	Spring stiffness	48 N/m

EXPERIMENTAL RESULTS AND DISCUSSION

Measurement setup

The fabricated prototype was tested using both a vibration exciter test. The vibration exciter test was carried out by mounting the prototype on vibration exciter as shown in Figure 6. Picoscope was used to measure output voltage from prototype.

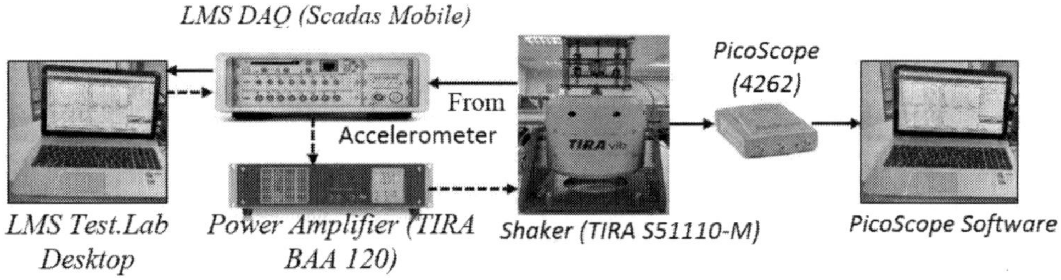

Figure 6. Schematic diagram of the vibration exciter test setup for the fabricated harvester prototype.

Experimental results and discussion.

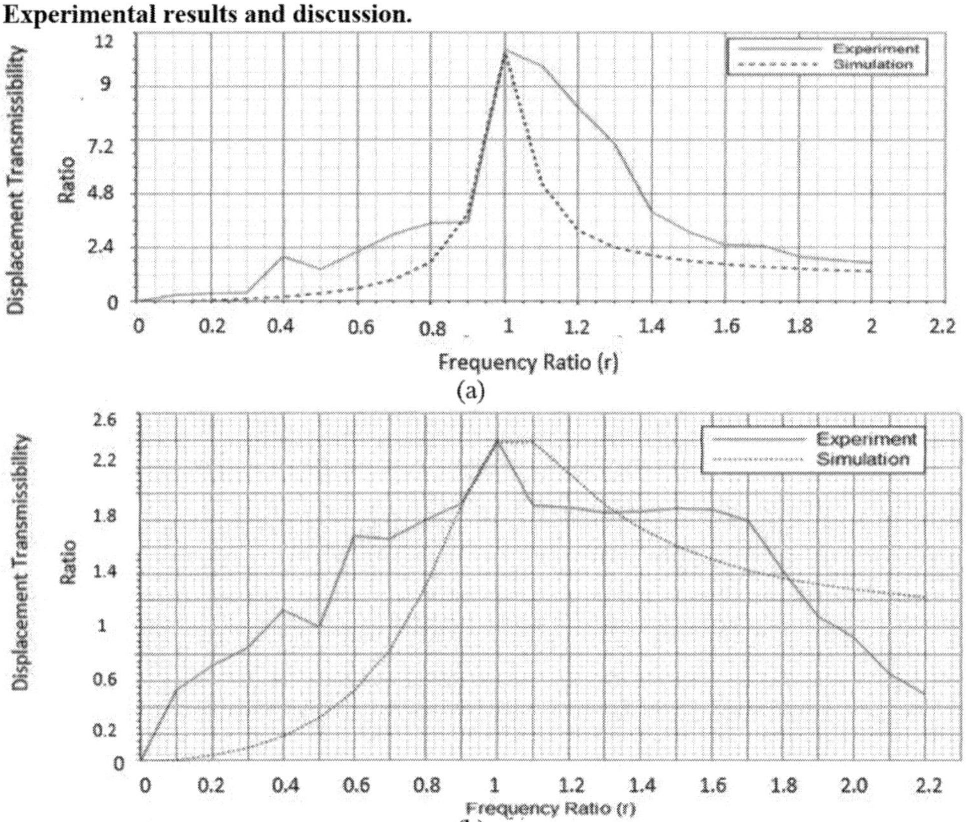

Figure 7. Comparison vibration transmissibility between experimental and simulation result for (a) upper and (b) lower system.

Based on the graph pattern in Figure 7, the experimental result for both systems have shown the almost same pattern flow with simulation result. Frequency ratio from 0 until 0.9 shows a little bit difference gap between experimental and simulation result where the experiment results have shown the higher responded for displacement transmissibility value compared with simulation result. The reason why this happened because the factors of low damping ratio and seismic mass. From the test results for upper and lower system have proved that with two moving mechanical systems can be improved functionality of the energy harvester by increase the bandwidth frequency range during operating especially using electromagnetic as transducer because almost all the electromagnetic transducer has two components, so moving both components can enhance the overall system function compared depending only one moving electromagnetic component during operating

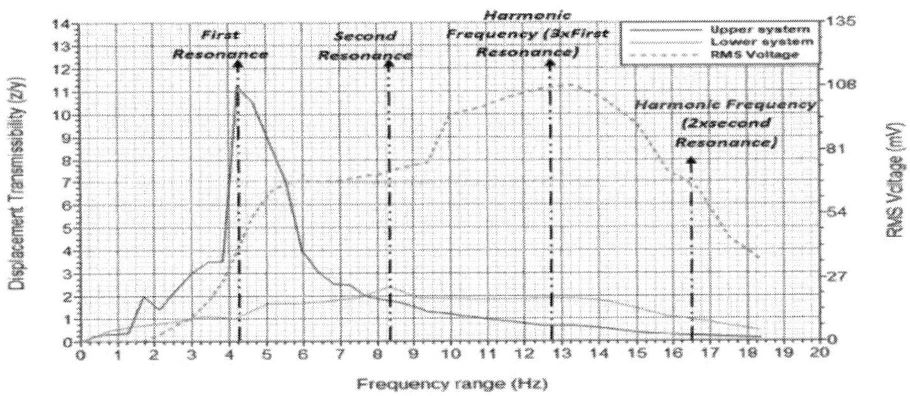

Figure 8. Output voltage result from the motion of upper and lower system at varying frequency with 0.04 g external amplitude value

Based in Figure 8, the effectiveness range of the operating frequency for energy harvester system was starting from 1.9 Hz until 18.5 Hz. At upper natural frequency 4.26 Hz, this system has capability to produce almost 54 mV even at low frequency. At this condition, the upper system has shown more displacement transmissibility about 11time bigger compare to lower, so the magnetic core had moved more across the copper coil. For the second resonance frequency at 8.34 Hz, the harvester system has produced output voltage about 80.52 mV which 50 % more than first resonance frequency 4.26 Hz.

Conclusion.

The research work has successfully developed an electromagnetic vibration energy harvesting system to operate at wide bandwidth frequency and low frequency. By introduced new technique which using two moving mechanical systems and two resonance frequency systems has successfully increased the output voltage of the harvester system. With controlling damping parameter value in mechanical system of energy harvester system, it has successfully made the system has capability to fully utilize the vibration level from external sources. The new mechanical mechanism concept that was applied in proposed energy harvester system has increased the effectiveness and efficiency of the vibration energy harvester system performance.

REFERENCES

[1] J. Twiefel and H. Westermann, "Survey on broadband techniques for vibration energy harvesting," *J. Intell. Mater. Syst. Struct.*, pp. 1291–1302.

[2] S. P. Beeby, M. J. Tudor, and N. M. White, "Energy harvesting vibration sources for microsystems applications," *Meas. Sci. Technol.*, vol. 17, no. 12, 2006.

[3] S. Cheng, C. Cepnik, and David P.Arnold, "Electrodynamic Vibrational Enegy Harvesting," in *Micro Energy Harvesting*, Weinheim, Germany: Wiley-VCH Verlag Gmbh & Co. kGaA, 2015, pp. 175–199.

[4] D. Spreemann and Y. Manoli, "Basic Analytical Tools for the Design of Resonant Vibration Transducers," in *Electromagnetic Vibration Energy Harvesting Devices*, 2012.

[5] R. L. Harne and K. W. Wang, "A review of the recent research on vibration energy harvesting via bistable systems," *Smart Mater. Struct.*, vol. 22, no. 2, 2013.

[6] A. R. M. Siddique, S. Mahmud, and B. Van Heyst, "Energy conversion by T-shaped cantilever type electromagnetic vibration based micro power generator from low frequency vibration sources," *Energy Convers. Manag.*, vol. 133, pp. 399–410, 2017.

[7] B. Lee and G. Chung, "Frequency tuning design for vibration-driven electromagnetic energy harvester," pp. 801–808, 2015.

[8] B. Yang *et al.*, "Electromagnetic energy harvesting from vibrations of multiple frequencies," vol. 19, 2009.

[9] C. B. Williams and R. B. Yates, "Analysis of a micro-electric generator for microsystems," *Science (80-.).*, vol. 52, pp. 8–11, 1996.

EXPERIMENTAL IDENTIFICATION ON NON LINEAR PROPERTIES OF RUBBER MOUNT

Iskandazaqwan Zikrullah Zainudin and Ooi Lu Ean

School of Mechanical Engineering,
Universiti Sains Malaysia
*Email: ooiluean@usm.my

ABSTRACT

In actual installation, rubber mount are usually excited by different engine force amplitude and frequency. For better characterization of rubber mount, dynamic properties of metal to rubber mount are tested in this paper by using hysteresis loop method. Stiffness and loss factor of rubber mount are calculated from measured hysteresis loop. Experimental works are carried to identify the non-linearity in the amplitude dependent and frequency dependent properties of rubber mount. The s-shaped of hysteresis loop represent as non-linear behavior of rubber mount. The comparison is done for the dynamic properties of rubber mount under different excitation condition. The non-linear behavior of the rubber mount under excitation forced are reported. The result show stiffness change non-linearly according to different amplitude excitation force under different excitation frequency. The observation is significant especially when the excitations force is higher than 5N for the small metal to the rubber mounts. However this observation is different compared to the condition where excitation frequency getting higher. The non-linearity in the rubber mount is becoming not significant when the excitation frequency is getting higher.

Keywords: Rubber mount, Stiffness, Force amplitude and Non-linearity

INTRODUCTION

Rubber mounts are commonly used to resolve vibration and noise in any device or machine since 1930[1]. Although several mounting system had been developed from elastomeric mount to hydraulic mount, from passive to active, the rubber mount still been selected to studied due to its compact, cost-effective, maintenance free and provided consistent performance[2][3]. In general, the performance of rubber mounts is influenced by dynamic properties of rubber components[4]. The dynamic properties of rubber mounts include stiffness and loss factor. Stiffness contributed to the energy store in system and damping characteristic is representing energy loss from the system.

Viscoelastic behavior of rubber make rubber mount to deform linearly and non-linearly under different excitation frequency and amplitude[5]. In order to obtain good description of viscoelastic material, several methods have been carried out. Shimizu et al. (1999) and Fukunaga et al. (2004) modeled viscoelastic materials by the fractional derivative model[6][7]. However, experimental validation was not been carried out during their studies. Lin et al. (2005) and Ooi et al. (2014) were conducted experiment study to evaluate frequency dependent stiffness and frequency dependent lost factor of rubber mount by using impact technique[8][9]. However, the authors were not focused on non-linear region small force excitation was applied in the studied and linear region behavior is assumed in the study. The non-linear frequency dependent properties of rubber mount were studied by Sjoberg et al. and Zhu et al. in both experiment and simulation[10][11]. However, authors only focused their studied in low frequency range which is in between 0 Hz to 100 Hz.

Gade et al. used resonant and non-resonant methods to study complex modulus and loss factor of rubber mount which is shaker is used as excitation source. However, the experiment carried out by authors focused on linear behavior of frequency dependent only[12]. Lin et al. was used impact technique to evaluate the frequency dependent stiffness and damping characteristic of rubber mount. However, the measured force and acceleration responses of the rubber mounts are determined by using Fast Fourier Transform (FFT)[8]. Ooi et al. (2011) further developed the impact technique to determine dynamic transfer stiffness and dynamic driving point stiffness. The difference between dynamic driving point stiffness and dynamic transfer stiffness due to the force applied and measured was considered at different position of the rubber mount[13]. Recently, shaker had been used in the experimental measurement in the low frequency testing on the cable isolators by Guzman et al. to study non-linear behavior of rubber mount. The damping properties are determined by using the shaker test[14]. The shaker excitation method is also applied by Vangipuram et al. in the dynamic characterization of rubber mounts in terms of linear and non-linear behavior.[15].

Hysteresis loop method can be used in several application such as rubber, magnetic field, electrical and so on[10][16][17]. Due to damping rubber material, hysteresis loop are obtained when plotting amplitude of instantaneous force versus instantaneous displacement in a material of all value of time. The linear behavior for dynamic properties of rubber mount will visualize the hysteresis loop in elliptical shape. For non-linear behavior, the shape of elliptical of hysteresis loop will be change to s-shape[18]. Hysteresis loop method is proposed here to provide alternative method in characterization of non-linear behavior of rubber mount.

The study presented in this paper will measure the frequency dependent and amplitude dependent stiffness of rubber mount based on hysteresis loop method. The non-linear behavior is determined within this frequency and amplitude dependent range. The reason to measure frequency dependent and amplitude dependent stiffness at several force and frequency is to study the effect of force and frequency of non-linear properties of rubber.

METHODOLOGY

Experimental measurement

The experiment setup used in this study consist of three rubber mounts, an accelerometer, force transducer, 0.9kg preload mass, power amplifier, LMS Scadas, stringer and shaker as the controlled excitation force. Rubber mount with diameter of 15mm and length of 20mm are used. Figure 1(a) shows the experiment setup and the apparatus used in this experiment. The different forces at different frequencies are set on Spectral Testing software and LMS Scadas. LMS Scadas is used as interface with controller to provide signal and collect data. The controller is used to generate voltage sinusoidal wave signal in order to drive the shaker as excitation source. The excitation force and frequency from shaker then will transmitted to the force transducer via stringer. The force transducer is used to measure the input force from the shaker. As Figure 1(b), the accelerometer located below the preload is used to measure the acceleration response of engine mounts. The experiment is repeated at differences force excitation of 6N and 10N. The experiment also will repeated with different excitation of frequency. First, the natural frequency of rubber mount is measured by determined the Frequency Response Function (FRF) of the system. Lin et. Al. stated in their studied, the range of natural frequency is determined by defined the half of FRF peak value as stated [8]. Then, the curved of FRF then subdivided into three section: (1) below natural frequency range; (2) within natural frequency range; (3) above natural frequency range.

Below natural frequency range ($0 \leqslant f \leqslant 90$ Hz).

In this experiment, the frequency used for below natural frequency range is 80 Hz.

Within natural frequency range ($90 \leqslant f \leqslant 140$ Hz)

Frequency of 100Hz is chose for study the stiffness of rubber mount within the natural frequency range

Above natural frequency range ($140 \leqslant f \leqslant 400$ Hz)

The frequency used for above natural frequency is 200 Hz. The above natural frequency range is ended at 300 Hz where the noise contamination in data becomes severe.

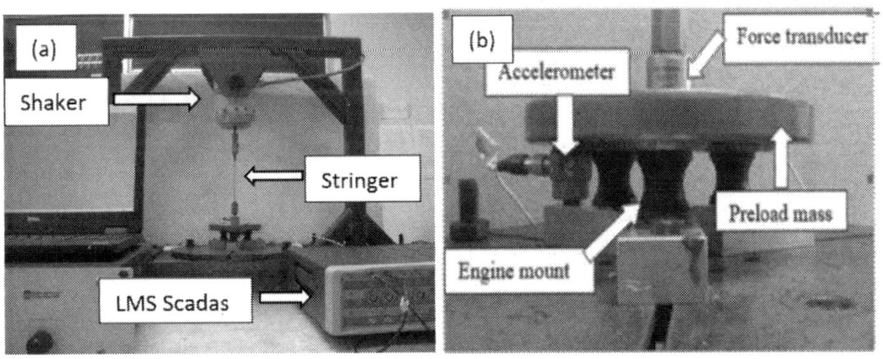

Figure 1. (a) Experiment setup for engine mount and (b) Experiment setup for sine swept method

Hysteresis loops are plotted from the data collected in above setup. Stiffness can be obtained by determine the slope of the loop[10][18]. From the Figure 2, the area of the loop represent as energy lost to the system. The stiffness can be calculated by determine the gradients of each force and displacement from the s-shaped loop. The graph is plotted to determine the stiffness at different gradient for each force and displacement of non-linear s-shaped loops.

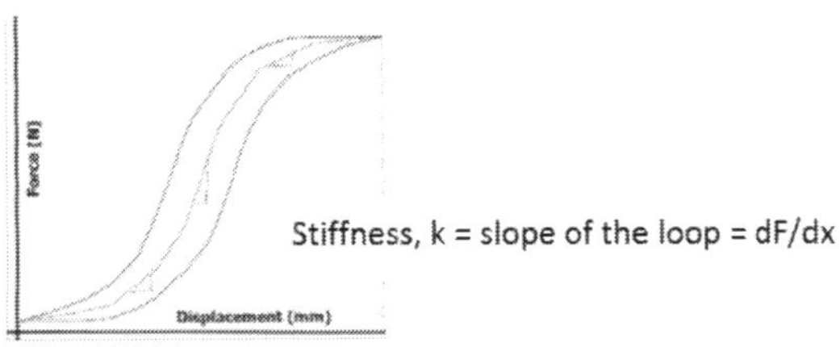

Figure 2: Scematic of Hysteresis loop of rubber mount

RESULTS AND DISCUSSION

The frequency response function (FRF) of the system is measured. Natural frequency is determined from the peak value of the FRF that is 113.5Hz. In this study, three frequencies have been chosen to represent three different natural frequency range. For below natural frequency, 80 Hz is selected. Within the natural frequency, 100 Hz is selected and for above natural frequency, 200 Hz is selcted.

Figure 3(a) shows the hysteresis loops that measured with different excitation force. The excitation force applied are 3N, 5N and 10N. From Figure 3(a), the shape of hysteresis loop is s-shape when the force is applied. The s-shape indicated the rubber mount behave non-linearly. Similar finding was reported by Aiken et. al that shows similar shape of hysteresis loops for non-linear behavior in their analytical studied of hysteresis model[19]. Stiffness was calculated as explained in methodology section. Figure 3(b) shows the stiffness calculated based on result from different excitation force but constant excitation frequency. From Figure 3(b), at excitation of 5N, the higher the force the higher the stiffness of rubber mount. The results of 7N and 10N excitation force shows similar trend with 5N. At 1N of 5N excitation force is 133.33kN/mm, while 1N for 7N and 10N of excitation forces are 109.8kN/mm and 90.9kN/mm respectively. The percentage different between 1N of 5N, 7N and 10N excitation force is less than 20%. Similar trend of stiffness shows from 2N to 5N at different of 5N, 7N and 10N excitation force.

Figure 3(c) shows the hysteresis loops that measured at 100Hz but different excitation force. The force applied were 3N, 6N and 7N. The results represented properties of rubber mount within natural frequency range. From Figure 3(c), the loops exhibit linear and the area of the loops are bigger compared to result obtained at 80Hz and 200Hz as shown from Figure 3(a) and Figure 3(e) respectively. Zainudin et al. (2017) also found the area of the loop become bigger when the excitation force near natural frequency of rubber since more energy loss to the system[20]. Figure 3(c) shows that the linear behavior occur to the system when excitation frequency within the natural frequency since the loops is elliptical shape. Figure 3(d) shows the stiffness of the rubber mount under different excitation force. The change of stiffness is very small when the excitation force change from 1N to 7N. The different is below t 15% for each force at different excitation force.

Figure 3(e) shows the hysteresis loops that measured at 200Hz but different excitation force. Same force level were applied for this frequency range. From the Figure 3(e) the shape of the hysteresis loops is elliptical shape which is linear behavior of the rubber mount system. Similar finding was reported by Ramorino et al. found the linear behavior of rubber mount when excited at higher than natural frequency. Figure 3(f) shows that the increase of the excitation force, the stiffness also increase. The different of stiffness between 3N, 6N and 7N of excitation force for 1N is smaller. The different forces of at 1N excitation force is below than 10%. The similar trend were found at different excitation force.

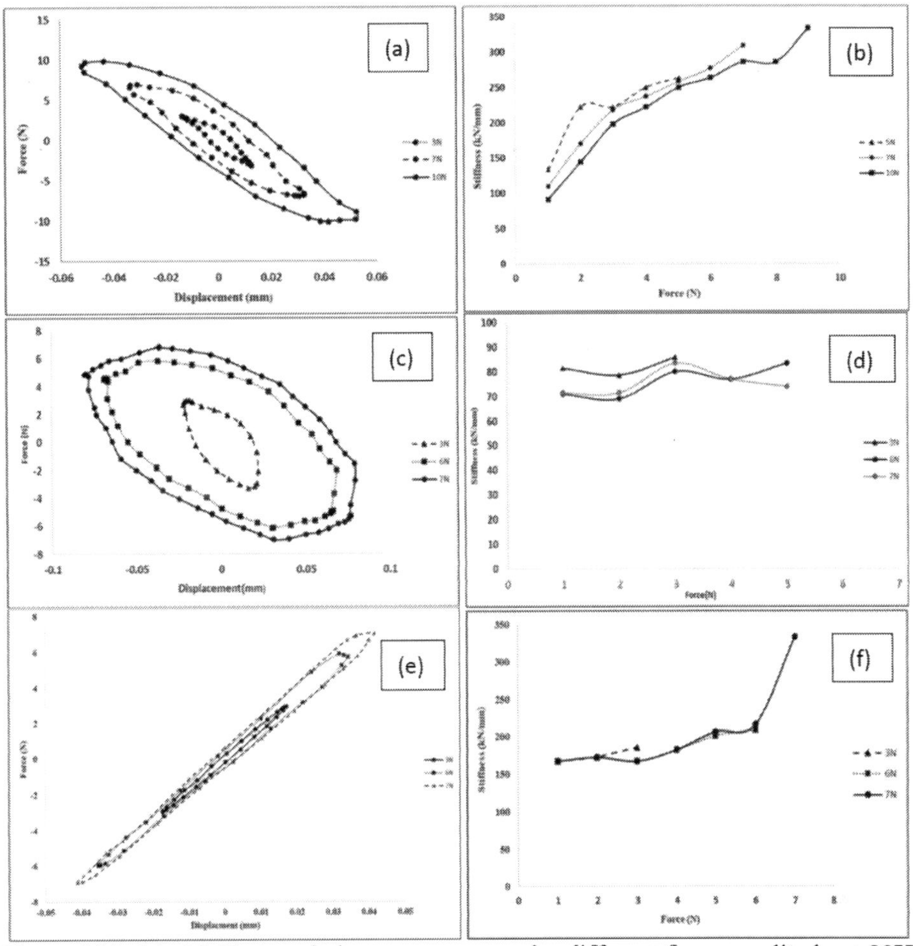

Figure 3(a), (c), (e) Hysteresis loop measures under different force amplitude at 80Hz, 100Hz, 200Hz. 3(b), (d), (f) Stiffness under different excitation force amplitude at 200H

The shape of loops at different frequencies are different. For the excitation frequency below and above natural frequency range, the shape of the loops are elliptical. It indicated the loops is linear. For the excitation frequency within natural frequency range, the shape of the loop is s-shape. It indicated that the loops is non-linear. Similar finding were reported by Aiken et. Al. where the shape of non-linear hysteresis loops is s-shaped. From figure 3(a), (c), (e) it is clearly seen that the shape of the hysteresis loop is not perfect. This might be cause by the external energy loss from rig and the connection of screw and nut between preload and rubber mount. The vibration through the rig and the screw and nut will cause energy loss and the losing in the thread of the screw significantly affect the input signal of the system [20].

CONCLUSION

The measurement using hysteresis loops method is carried out for different frequency and excitation force. The results shows stiffness change non-linearly when excited within natural frequency range. The observation is significant when the excitation force is higher than 5N. However this observation is different compared to the condition where excitation frequency is excited below and above natural frequency range.

ACKNOWLEDGEMENTS

The author would like to gratefully acknowledge Univsersiti Sains Malaysia (USM) Short Term Grant [account no. 60313028] and Ministry of Higher Education Fundamental Research Grant Scheme [account no. 6071345] for funding the project.

REFERENCES

[1] H. C. Lord, "*Vibration-dampening mounting*", *United States Patern Office,* no. 1778503. 1930.

[2] R. V Schmitt and C. J. Leingang, "Design of Elastomeric Vibration Isolation Mounting Systems for Internal Combustion Engines." *SAE Technical Paper 760431*, 1976.

[3] M. W. Sayers and C. Mink, "A Simulation Graphical User Interface for Vehicle Dynamics Models," *SAE Trans.*, vol. 104, pp. 58–67, 1995.

[4] Y. Yu, N. G. Naganathan, and R. V Dukkipati, "A literature review of automotive vehicle engine mounting systems," *Mechanism and Machine Theory*, vol. 36, pp. 123–142, 2001.

[5] N. Micali and M. Carlo, "Viscoelastic properties," *Physical Review A*, vol. 42, no. 12, 1990.

[6] W. Shimizu, N., & Zhang, "Fractional calculus approach to dynamic problems of viscoelastic materials," *JSME Int.*, vol. 4, no. 42, pp. 825–837, 1999.

[7] N. Fukunaga, M., & Shimizu, "Analytical and numerical solutions for fractional viscoelastic equations," *JSME Int.*, vol. 47, no. 1, pp. 251–259, 2004.

[8] J. Lin, T. R., Farag, N. H., & Pan, "Evaluation of frequency dependent rubber mount stiffness and damping by impact test.," *Appl. Acoust.*, vol. 7, no. 66, pp. 829–844, 2005.

[9] Ooi L. E. and Z. M. Ripin, "Optimization of an engine mounting system with consideration of frequency-dependent stiffness and loss factor," *J. Vib. Control*, vol. 22, no. 10, pp. 2406–2419, 2014.

[10] M. M. Sjoberg and L. Kari, "Non-Linear Behavior of a Rubber Isolator System Using Fractional Derivatives," *Veh. Syst. Dyn.*, vol. 37, no. 3, pp. 217–236, 2002.

[11] P. D. Zhu, S., Cai, C., & Spanos, "A nonlinear and fractional derivative viscoelastic model for rail pads in the dynamic analysis of coupled vehicle–slab track systems.," *J. Sound Vib.*, no. 335, pp. 304–320, 2015.

[12] S. Gade, H. Herlufsen, K. Zaveri, and K.-H. H., "Damping Measurements - From Impulse Response Functions - From Resonance and Non-resonance Excitation Techniques," *Brüel Kjaer Tech. Rev.*, vol. 2, no. 2, p. 50, 1994.

[13] L. E. Ooi and Z. M. Ripin, "Dynamic stiffness and loss factor measurement of engine rubber mount by impact test," *Mater. Des.*, vol. 32, no. 4, pp. 1880–1887, 2011.

[14] D. F. Guzmán-Nieto, M., Tapia-González, P. E., & Ledezma-Ramírez, "Low frequency experimental analysis of dry friction damping in cable isolators.," *J. Low Freq. Noise, Vib. Act. Control*, vol. 4, no. 34, pp. 513–524, 2015.

[15] P. Vangipuram, R., Padmanablan, C., & Ravindran, "Dynamic characterization of rubber mounts.," *23rd Int. Congr. Sound Vib.*, pp. 10–14, 2016.

[16] D. Bachleitner-Hofmann, A., Abert, C., Bruckner, F., Palmesi, P., Satz, A., & Suess, "Unexpected width of minor magnetic hysteresis loops in nanostructures," *IEEE Trans. Magn.*, vol. 7, no. 52, pp. 1–4, 2016.

[17] and D. Y. Wei, Jing, Yicheng Zhao, Heng Li, Guobao Li, Jinlong Pan, Dongsheng Xu, Qing Zhao, "Hysteresis analysis based on the ferroelectric effect in hybrid perovskite solar cells.," *J. Phys. Chem. Lett.*, vol. 5, no. 21, pp. 3937–3945, 2014.

[18] J. P. Nashif, A. D., Jones, D. I., & Henderson, *Vibration damping. A Wiley-Interscience Publication*, 1985.

[19] I. D. Kikuchi, M., & Aiken, "An analytical hysteresis model for elastomeric seismic isolation bearings," *Earthq. Eng. Struct. Dyn.*, vol. 2, no. 26, pp. 215–231, 1997.

[20] I. Z. Zainudin, H. A. Darun, and O. L. Ean, "Dynamic Measurement of Engine Mount Properties Using Hysteresis Loop Method," no. August, 2017.

NUMERICAL AND EXPERIMENTAL INVESTIGATIONS OF NONLINEARITY BEHAVIOUR IN A SLENDER CANTILEVER BEAM

A. R. Bahari[1], M. A. Yunus[1,*], M. N. Abdul Rani[1], Ayub, M. A.[1], A. Nalisa[2]

[1]Structural Dynamics Analysis and Validation (SDAV),
Faculty of Mechanical Engineering,
Universiti Teknologi MARA, 40450 Shah Alam, Selangor, Malaysia
[2]Faculty of Mechanical Engineering, Universiti Teknologi MARA, Cawangan Johor,
Kampus Pasir Gudang, Bandar Seri Alam, 81750 Masai, Johor, Malaysia
*Email: mayunus@salam.uitm.edu.my
Phone: +60355436202

ABSTRACT

Nonlinear problem is always occur in slender structures that are usually characterized by large displacements and rotations but small strains. Linear design assumption could lead to premature failure if the structure behaves nonlinearly. In this paper, the static displacement of a slender beam subjected to point load is investigated numerically by incorporating the large amplitude of the displacement. Two types of numerical analyses are performed at a full-scale finite element model which is linear static and geometric nonlinear implicit static. The results of the FEA linear static analysis are compared with the results from the FEA geometric nonlinear implicit static analysis. It shows that very high different load-displacement value response. Experimental static displacement test has been performed to validate both numerical results.

Keywords: geometrically nonlinear, structural analysis and linear static.

INTRODUCTION

Current designs and models in industry normally remain linear. These structural models are normally constructed based on idealized engineering designs and they may not accurately represent all the aspects of actual structures [1]. In some cases, linear models may fail to predict the structural response within the necessary level of reliability. Linear design could lead to premature failure if the structure behaves nonlinearly. Most of cases were assumed as a linear [2]

Nonlinearities in structures can lead to large amplitude responses due to extreme mechanical and environmental loading conditions. The investigation on the nonlinear behavior is very important because it effects to the integrity of the engineering structure. Two types of nonlinearities commonly encountered in structural analysis are geometric and material. Material nonlinearity is associated with nonlinear stress–strain relations whereas non-linear curvature-slope and strain-displacement relations give rise to geometric nonlinearity [3].

The researchers try to consider nonlinear phenomena for structures and devices. Some examples of nonlinear mechanical systems phenomena occur in aerospace sector, guyed tower and rotating wind turbine blade [4, 5]. In aeronautic, a new aircraft designs with light weight and high aspect-ratio wing provides advantage in producing a higher lift-to-drag ratio [6]. These result the wing to be very flexible and cause large deformations under normal operating loads leading to a geometrical nonlinear behavior and aeroelastic problems [7-9].

Geometrically non-linear problem is always occur at slender structures that are usually characterized by large displacements but small strains [10] and relates with the geometric effect. Large deflection problem of cantilever beams is generally analyzed in curvilinear coordinate system. Euler Bernoulli beam theory in curvilinear coordinate system (s,n) is $1/\rho = M/EI$, where curvature $1/\rho = d\varphi/ds$. So Euler Bernoulli bending moment-curvature relationship is given as follows [3] and φ is the slope dy/dx at location s, and it is also the measure of normal direction n.

$$EI\frac{d\varphi}{ds} = M$$

The objective of this paper is to apply a procedure for investigating the large deflection of a straight cantilever beam subjected to a tip concentrated load. In the present work, we demonstrate the separation between two methods, one is linear static analysis and the other one is geometrically nonlinear implicit static analysis. The paper is organized as follows. Section 1 briefly introduces the geometric nonlinearity in structural modelling. Section 2 describes geometrical and structural features used for experimental and numerical works. Section 3 presents a result of the experimental and numerical analysis performed on the sample. Finally, Section 4 concludes the paper.

METHODOLOGY

Experimental Setup

A static load experiment has been specifically designed and performed in laboratory in order to investigate geometric nonlinearity in structural analysis. The experimental set-up consists of a slender beam (standard ruler) with the dimension of 1024 mm × 32 mm × 1.5 mm (length × width × thickness) and a hole of diameter 7.3 mm at one end. The beam was clamped at the left end side by using two G-clamps and the boundary condition is fixed support. This assembly is illustrated in Figure 1.

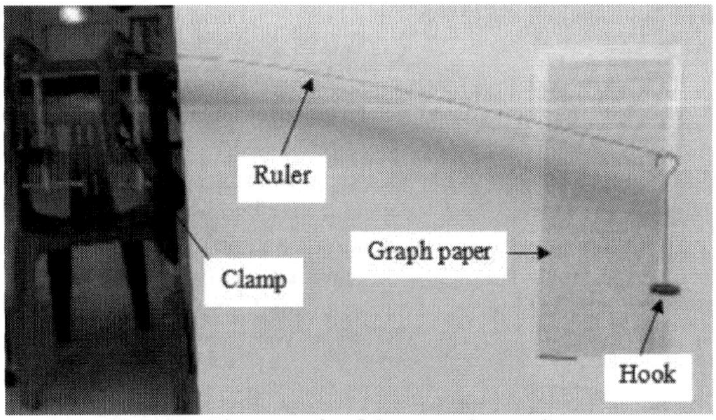

Figure 1. Photograph of the experimental setup

In this work, a graph paper is placed on the wall in order to accurately measure the displacement data. The ruler is point loaded at the right end side by placing several defined weights on the weight hook. The deflection of the ruler for each weight was measured through the marking point on the graph paper. In each step of loading, the deflection profile is captured and recorded manually.

Finite Element Model and Analysis

The finite element of the beam was developed using Hypermesh with 3813 elements and 8350 nodes. The physical properties used in the finite element modeling are elastic modulus E = 193000 GPa, density ρ = 7700 kg/m^3 and Poisson's ratio ν = 0.33. Geometric details and ruler model developed for the present study are shown in Figure 2.

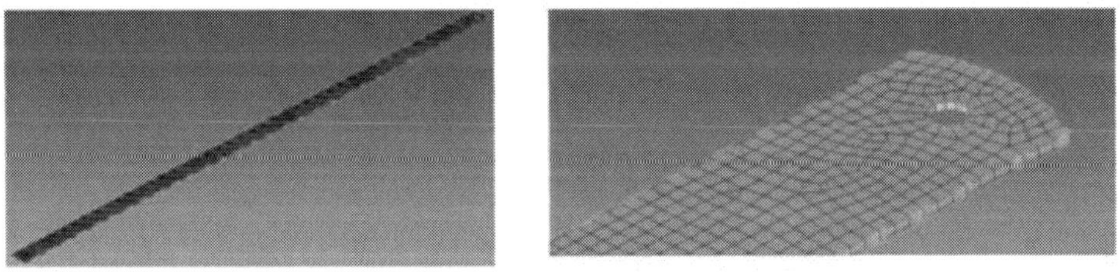

(a) (b)

Figure 2. A ruler model (a) isometric view of the CAD model (b) meshed model

The linear and nonlinear effects have been simulated via displacement-based finite element analysis results. The fixed support at one end in the experiment was modelled by fully constrained based on a single point constraint. Eleven sets of different concentrated point load have been assigned to the end of ruler. The large displacement was computed from this model by using two types of loadsteps: linear static analysis and geometrical nonlinear implicit static analysis.

RESULTS AND DISCUSSION

Experimental Deflection Profiles

Figure 3 shows the deflection of the cantilever beam for several weight point loads. The deflection comes from the weight of the ruler (distributed load) and weight from the blocks (concentrated load). In zero load condition, the ruler already deflects by exact value due to the weight itself. The experimental data appear to have large deflection during the low value of the applied load due to the geometrically nonlinear behavior of the structure.

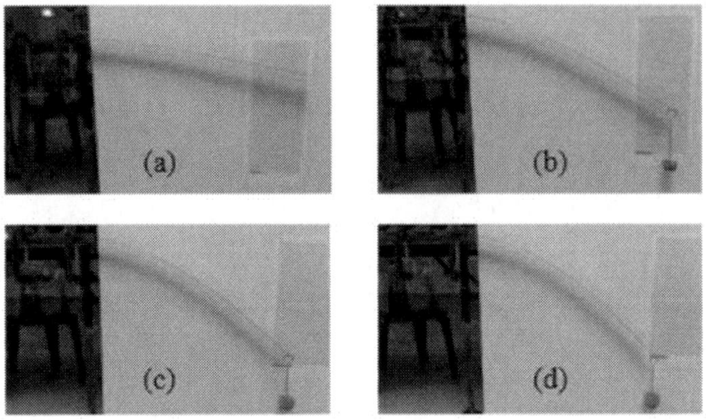

Figure 3. Experimental deflection of a ruler for point load (a) zero (b) 2.206 N (c) 4.206 N (d) 5.206 N

Comparison between Experimental and Numerical Analysis Results

The deflection of the cantilever beam by numerical analysis for 4.206 N point load using linear static and geometric nonlinear implicit static analyses is shown in Figure 4. Table 1 shows the detail value of the deflection for each of point load by experiment and finite element analysis using linear static and geometric nonlinear implicit static analyses. To perform detail investigation of the deflection, a plot of the displacement values in Table 1 is given in Figure 5.

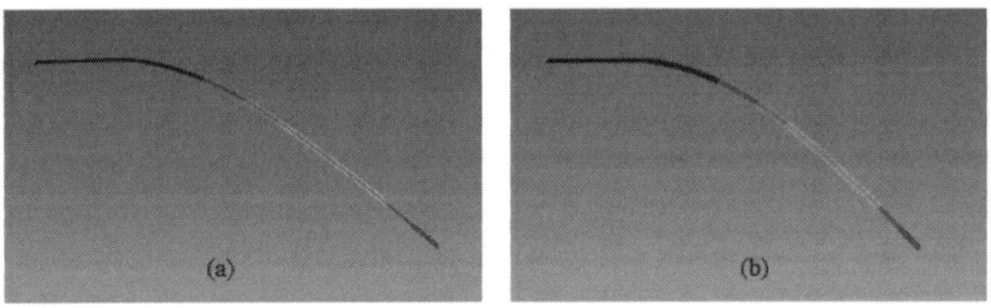

Figure 4. Numerical analysis deflection of a ruler using (a) linear static and (b) geometric nonlinear implicit static analyses

Table 1. Deflection values from linear static and geometric nonlinear implicit static analyses.

	Experiment (Geometric Nonlinear)	FEA (Linear Static)	Error	FEA (Geometric Nonlinear)	Error
Load (N)	Displacement (mm)	Displacement (mm)	%	Displacement (mm)	%
0	126	120	4.6	119	5.9
0.206	153	142	7.2	140	8.8
0.706	197	195	1.0	188	4.4
1.206	243	248	2.0	234	3.7
1.706	282	301	6.7	277	1.9
2.206	316	354	12.0	316	0.0
2.706	350	407	16.3	351	0.5
3.206	379	460	21.4	384	1.5
3.706	402	512	27.6	414	3.1
4.206	427	566	32.6	442	3.5
4.706	447	619	38.6	467	4.5
5.206	464	672	45.0	490	5.6

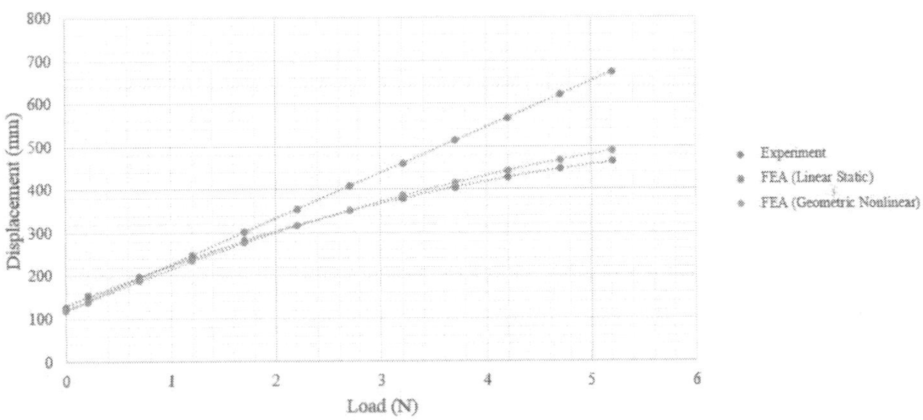

Figure 5. Deflection of a ruler using linear static and geometric nonlinear implicit static

Initial displacement measured is 126 mm experimentally in its free state (weight of ruler only) condition. By using numerically linear static analysis, the displacement is 120 mm. It is observed that this method produces constant increment of displacement values when the point load increases. The increment pattern is consistent with linear characteristic response. The error also increases since the discrepancy between experiment and linear static analysis value increase.

A good agreement of numerical and experimental results has been found for the geometric nonlinear implicit static analysis. Due to the long slender ruler with thin property, the deflection imposed geometric nonlinearity. It is obvious from Figure 5, it can be seen that the geometric nonlinear implicit static analysis shows less discrepancies with real experimental structure and give more precise predictions of geometric nonlinear behavior for future analyses. The best fit linear load-deflection curve is obtained from the calculated data points and the geometric nonlinear is clearly visible. Hence, this large deflection tip load is considered to capture the effect of geometric nonlinear in actual case.

CONCLUSION

This paper discusses the deflection behavior of the cantilever beam subjected to the concentrated load. Experiment has been carried out at the long thin stainless steel ruler to measure its deflection. Next, the deflection was calculated by performing numerical modelling and analysis based on two types of static analysis: linear static and geometric nonlinear implicit static. Finally the results were compared and successfully judge the geometric nonlinear deflection behavior in a beam.

REFERENCES

[1] Wang, Z.-C., Xin, Y. and Ren, W.-X. (2015). Nonlinear structural model updating based on instantaneous frequencies and amplitudes of the decomposed dynamic responses, *Engineering Structures*, 100: 189–200.

[2] Sulaiman, M. S. A., Yunus, M. A., Bahari, A. R. and Abdul Rani, M. N. (2017). Identification of damage based on frequency response function (FRF) data. *MATEC Web of Conferences*, 90, 01025.

[3] Ghuku, S. and Saha, K. N. (2016). A theoretical and experimental study on geometric nonlinearity of initially curved cantilever beams, *Engineering Science and Technology, an International Journal*, 19: 135–146.

[4] Overgaard, L. C. T., Lund, E. and Camanho, P. P. (2010). A methodology for the structural analysis of composite wind turbine blades under geometric and material induced instabilities, *Computers and Structures*, 88: 1092–1109.

[5] Eder, M. A., Bitsche, R. D. and Belloni, F. (2015). Effects of geometric non-linearity on energy release rates in a realistic wind turbine blade cross section, *Composite Structures*, 132: 1075-1084.

[6] Yunus, M. A., Abdul Rani, M. N., Sani, M. S. M. and Aziz Shah, M. A. S. (2018). Finite element model updating of riveted joints of simplified model aircraft structure. *AIP Conference Proceedings*, 1952, 020013.

[7] Spada, C., Afonso, F., Lau, F. and Suleman, A. (2017). Nonlinear aeroelastic scaling of high aspect-ratio wings, *Aerospace Science and Technology*, 63: 363-371.

[8] Changchuan, X., Chao, A., Yi, L. and Chao, Y. (2017). Static aeroelastic analysis including geometric nonlinearities based on reduced order model, *Chinese Journal of Aeronautics*, 30(2): 638–650.

[9] Tang, D. and Dowell, E. H. (2002). Experimental and Theoretical Study of Gust Response for High-Aspect-Ratio Wing, *American Institute of Aeronautics and Astronautics Journal*, 40(3): 419-429.

[10] Mian, H-H, Wang, G. and Ye, Z-Y. (2014). Numerical investigation of structural geometric nonlinearity effect in high-aspect-ratio wing using CFD/CSD coupled approach, *Journal of Fluids and Structures*, 49: 186-201.

BREATHING SOUNDS – DETERMINATION OF EXTREMELY LOW SPL

M. Harun[1*], R. Teoh Y. S[1], M. 'A. A. Ahmad, M. Mohd. Mokji[1], You K. Y[1], S. A. R. Syed Abu Bakar[1], P. I. Khalid[1], S. Z. Abd. Hamid[1] and R. Arsat[1]

[1] School of Electrical Engineering,
Universiti Teknologi Malaysia,
*Email: mokhtarharun@utm.my
Phone: +6075535358

ABSTRACT

Breathing sound is an extremely low SPL that results from inspiration and expiration process in the lung. Breathing sound can be used to diagnose persons with complications with breathing. Also, the sound can indicate the effectiveness of treatment of lung disease such as asthma. The purpose of this study was to identify SPL of breathing sounds, over six one octave center frequencies from 63 Hz to 4000 Hz, from the recorded breathing sounds in *.wav* files. Breathing sounds of twenty participants with normal weight BMI had been recorded in an audiometry room. The breathing sound was acquired in two states: at rest and after a 300 meters walk. Matlab had been used to process the breathing sounds that are in *.wav* files to come up with SPL (in dB). It has been found out that the SPL of breathing sound of all participants are positive at frequencies 63 Hz and 125 Hz. On the other hand, the SPL are all negatives at frequency 1000 Hz, 2000 Hz and 4000 Hz. In conclusion, SPL of breathing sounds of the participants, at frequencies 250 Hz and 500 Hz that have both positive and negative values are viable to be studied further for physiological and medicinal clues.

Keywords: Sound pressure, extremely low SPL, respiratory, breathing sound.

INTRODUCTION

Measurement of the extremely low sound pressure level (SPL) is important to evaluate acoustics environment such as audiometry room and recording studios. Noise emitted from quiet machinery and equipment such as lighting armatures is also part of extremely low SPL [1]. Elements in extremely low SPL in human such as breathing sound and heart sound are important to determine characteristics of physiological qualities for monitoring of organ functioning, effectiveness of respiratory treatment, and detection of disease [2][3].

By using compact and conventional sound level meter (SLM), direct measurement of extremely low SPL using sound level meter is very challenging due to two factors. First, the reference sound pressure which is 20 μPa used in the meter is the lowest sound pressure the machine can detect. Then, the meter will give the reading 0 dB. If the impinging sound pressure on the microphone is lower than 20 μPa, these sounds will not show up on the display on the meter. Second, the inherent noise in the SLM adds up the level of the noise is to be measured. Even if the noise level is known and compensated for, a satisfying measurement can only be done for levels 2-5 dB above noise floor [4].

The low frequency SPL causes annoyance even if it is below the limiting value, typically, 30 dB. For the measurement of low frequency extremely low SPL, the operational high pass filter has to be set to 20 Hz, so as not to overload the meter due to probable high sound pressure of low frequency noise input (Bjor, 1999). The measurement of low SPL at low frequency incurs high bias error and large scatter, Bias error indicates how the data deviate from the data average value. Large scatter in the data set will cause poor reproducibility. From the measurement of low sound pressure levels at low frequencies conducted in an Inter Nordic Round Robin comparison with 5 laboratories, it has been found out that the low frequency noise level has poor reproducibility up until frequency 200 Hz. The reproducibility is in the higher order of 15 dB. At frequency 250 Hz and higher, the bias error has gone down and reproducibility has improved satisfactorily. The findings also assert that noise due to handling of measurement instrument must be 6 dB below the measurement signals [4].

Breathing is also called ventilation or respiration is an act that consists of two breaths. One breath that is called inspiration is the process to take air into the lungs. The other breath, which is called expiration, is the process to take air out of the lungs. Ineffective ventilation can negatively impact efficacious gas exchange. For an adult with healthy lungs (or non-respiratory compromised), number of breathing should be normal between 12-20 number of breath per minute [[5] and [6], as cited in [3]]. In hospital, respiratory rates are used as the early warning scores (EWS) vital signs. When the breathing rate slows down to 8 breaths per minute or less, a patient is about 18 times higher odds of death within 24 hours when compared to a patient with a normal respiratory rate [[7], as cited in [3]].

Breathing sounds are extremely low SPL that occurs mostly at low frequency (Figure 1). A woman has breathing sounds at higher frequency compared to man [9]. Breathing complication sound such as wheezing sounds have pitch at 100 Hz and 1000 Hz. Rhonchi sounds, which is similar to wheezes but coarser has pitch around 150 Hz [8].

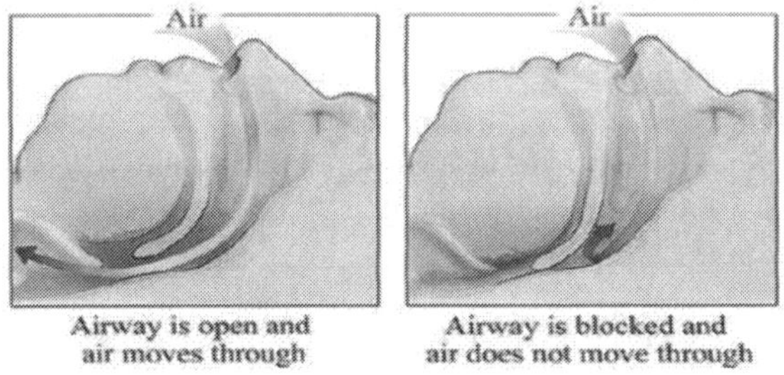

Figure 1. Breathing process: Normal (left); Obstructive breathing (right) [8]

METHODOLOGY

Experimental Setup

Twenty fourth-year students from School of Electrical Engineering (SEE) UTM Skudai had participated in the study. They were 10 males and 10 females. They shared the common BMI which is normal weight, in the range of BMI 20-25. The students were given specified time and date to come to the laboratory for the study. Audiometry room in Acoustics Research Laboratory, SEE UTM Skudai had been used as recording room. RealWave sound recorder that comes with 50 mV/Pa sensitivity microphone had been used to record the breathing sounds. The breathing sounds from each participant were recorded in two states: at rest and after 300 meters walk. For breathing sound recording, at rest, the participants was let resting in the lab not fewer than 5 minutes before the breathing sound was recorded. Then, the participants were asked to walk about 300 meters along the corridors of school building. After the participant finished the walk, breathing sound of the participant was recorded once again. The recorded breathing sounds (in *.wav* files) were compiled in terms of participants ID, gender, and breathing states: Rests or Walk. Matlab had been used to analyse these *.wav* files to output SPL at their respective frequency.

RESULTS AND DISCUSSION

Table 1 show SPL at one-octave frequency 63 Hz to 400 Hz breathing sounds for male participant at rest. SPLs at one-octave frequency 63 Hz and 125 Hz are all positives indicating that the sound pressures at these two frequencies are all above the reference pressure of 20 μPa. Beginning at frequency 250 Hz up until 4000 Hz, more SPL of breathing sound that is below 20 μPa has been found from the participants' recorded breathing sounds. At frequencies 2000 Hz and 4000 Hz, the measured SPL of breathing sound of male participants at rest are all negatives (sound pressure below 20 μPa).

Table 1. SPL (in dB) versus frequency (Hz) of sampled data - Male (Rest).

Participant/ Frequency (Hz)	63	125	250	500	1000	2000	4000
1	25	29	16	1	-13	-11	-13
2	26	26	21	9	-5	-10	-3
3	26	26	14	7	-10	-2	0
4	23	12	0	-4	-9	-6	-7
5	28	10	9	0	-14	-7	-9
6	35	28	20	7	-5	-14	-1
7	8	19	-1	0	-12	-19	-17
8	30	15	6	6	1	-12	-5
9	31	19	21	-1	-5	-10	-10
10	15	19	18	5	-27	-11	-12

It is apparent from Figure 2 that the SPL of breathing sound conforms to the conventional equal loudness contour. If one to compare SPL value of any soft physiological sound from human such as snoring and sneezing, the difference of breathing sound is that its SPL are all negatives at frequency 2000 Hz and 4000 Hz.

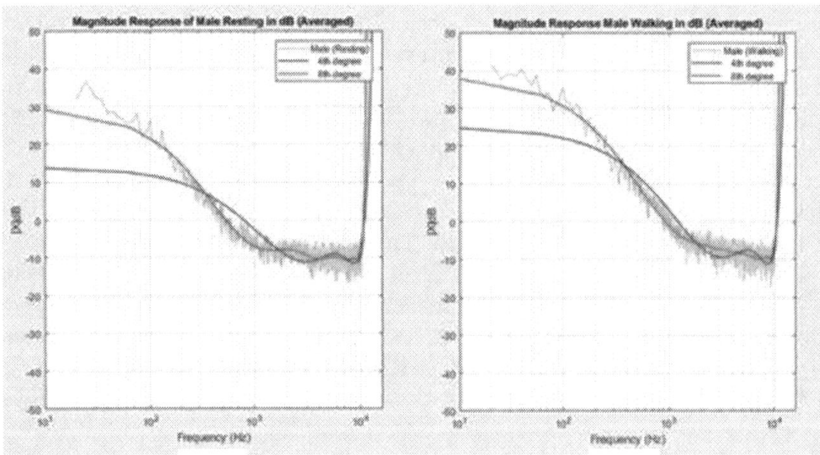

(a) (b)

Figure 1: SPL (dB) versus frequency (Hz) for male: (a) Rest, (b) Walk.

Statistical tests of the SPL results.

From Table 2, it is found that the mean values of SPL of breathing sounds after a walk are generally higher at all frequencies for both male and female participants. The values of standard deviation are also consistent of the increase for both male and female participants. Therefore, the data confirms to the hypothesis that SPL of breathing sounds is higher when there is physical activities of the participants, and thus the rate of metabolism.

However, it is found (in Table 3) that the variance of the SPL data across all frequencies for female participants is quite large. It can be inferred that there exist many large SPL or outliers in the data of the ten female participants. Also in SPL data of female participants, except for the frequencies 250 Hz and 1000 Hz, variances of the SPL of breathing sound is doubled for the Walk compared to the Rest.

Finally, from statistical significance tests between the Rest and the Walk, it is found that the SPL breathing sound for female participants is significance at frequencies 63 Hz, 125 Hz, and 1000 Hz.

Table 2. Statistical tests of SPL (dB) - Descriptive

Breath. state	Gender/ Freq. (Hz)	63	125	250	500	1000	2000	4000	Tests
Rest (Walk)	M	25(36)	20(30)	12(20)	3(9)	-10(-2)	-10(-6)	-8(-6)	Mean
	F	28(33)	25(34)	12(21)	3(10)	-4(3)	-9(-5)	-13(-9)	
	M	26(36)	19(33)	15(19)	3(10)	-9(0)	-10(-7)	-8(-5)	Med.
	F	25(31)	23(32)	9(24)	2(8)	-5(-2)	-8(-8)	-13(-7)	
	M	63(166)	46(126)	71(129)	18(134)	56(57)	22(69)	29(51)	Var.
	F	72(294)	114(216)	143(163)	115(331)	66(163)	24(165)	32(12)	
	M	8(13)	7(11)	8(11)	4(12)	7(8)	5(8)	5(7)	Std. dev
	F	8(17)	11(15)	12(13)	11(18)	8(13)	5(13)	6(3)	

Note: M denotes male, and F is for female

Table 3. Statistical test of SPL (dB) - Significance

Test of significance	Gender/ Frequency (Hz)	63	125	250	500	1000	2000	4000
Significance ($p<0.05$)	M	0.41	0.13	0.12	0.34	0.14	0.35	0.05
	F	0.04	0.03	0.12	0.15	0.04	0.22	0.68
Verdict	M	No	No	No	No	No	No	No
	F	Yes	Yes	No	No	Yes	No	No

Note: M denotes male, and F is for female

CONCLUSION

From the study, it can be concluded that SPL data at 250 Hz and 500 Hz during Rest and Walk breathing states for both genders can be investigated further for clues because at these frequencies there exist both positive and negative SPLs. High difference of variance of SPL data between the Rest and the Walk for female participants has inferred that the data need to be investigated for the influence of noise. If the influence of noise can be ruled out, the acquired data might be able to shed light on SPL of breathing sound of women with respect to physical activities. Finally, it can be concluded that physical activities of participants with common BMI affect SPL of their breathing sounds at certain frequencies.

ACKNOWLEDGEMENTS

The authors would like to acknowledge Universiti Teknologi Malaysia and Kementerian Pengajian Tinggi for providing the financial assistance under research grants FRGS Vot 4F870 and GUP Vot 14J47.

REFERENCES

[1] Bjor, Ole-H. (1999). Application note: Measurement of Extremely Low Sound Pressure Levels Ole-Herman Bjor, M.Sc. Norsonic AS. AN Low-noise mic. Ed. 1. Rev. 0 ENGLISH 04.99.

[2] Lozano-Garcia, M., Fiz, J. A., Martinez-Rivera, C., Torrents, A., Ruiz-Manzano, J., Jane, Â. R. (2017). Novel approach to continuous adventitious respiratory sound analysis for the assessment of bronchodilator response. PLoS ONE 12(2): e0171455. doi:10.1371/journal.pone.0171455.

[3] Flenady, T., Dwyer, T., Applegarth, J. (2017). Accurate respiratory rates count: So should you! Australasian Emergency Nursing Journal. Vol. 20 (2017), 45-47.

[4] Simmons, C. (2009). Measurement of sound pressure levels at low frequencies in rooms. NT Technical Report. TR385 (1997). Nordic Innovation Center (Norden). Oslo, Norway.

[5] Walsh M, Erb G, Kozier B. (2010). Fundamentals of Nursing: Frenchs Forest. N.S.W: Pearson Education; 21.

[6] McCance K. L, Huether S. E. (2015). Pathophysiology The biologic basis for disease in adults and children. Elsevier Health Sciences.

[7] Ljunggren M, Castrén M, Nordberg M, Kurland L. (2016). The association between vital signs and mortality in a retrospective cohort study of an unselected emergency department population. Scand J Trauma Resusc Emerg Med. Vol. 24(1): 21-.

[8] SASA - Scottish Association of Sleep Apnea. (2018). Website: https://scottishsleepapnoea.co.uk/patient-information/symptoms-diagnosis-and-treatment-of-sleep-apnoea/

[9] Gavriely, N., Nissan, M., Rubin, A. H., Cugell, D. W. (1995). Spectral characteristics of chest wall breathe sounds in normal subjects. Thorax 1995 50: 1292-1300. doi: 10.1136/thx.50.12.1292

MEASUREMENT OF THE HISSING-TYPE NOISE AND VIBRATION OF THE AUTOMOTIVE HVAC SYSTEM

Muhammad Abdul Rahman Paiman[1,2], Ahmad Zhafran Ahmad Mazlan[1*], Muhammad Hidayat Hamdan[1], Mohamad Syazwan Md. Isa[1], Sharum Man[2], Mohd Zukhairi Abd. Ghapar[2], Mohamad Shamsul Azraf Sulaiman[2] and Tomoyuki Miyashita[3]

[1]School of Mechanical Engineering, Engineering Campus, Universiti Sains Malaysia, Penang
[2]Vehicle Comfort, Testing & Development, Vehicle Development & Engineering, Proton Holdings Berhad, Selangor
[3]School of Creative Science and Engineering, Waseda University.
*Email: zhafran@usm.my

ABSTRACT

Noises such as hissing, humming, air-rush and compressor engagement are the common type of noises that can be induced from the automotive heating and ventilating air conditional (HVAC) system. These noises are basically generated from the effects of vibrational HVAC components. Due to this, the root cause of the noises has to be investigated for any implementation of solution. In this study, the hissing-type of noise is taken into consideration whereby the noise and vibration are measured from various HVAC components such as Evaporator Inlet and Outlet and Thermal Expansion Valve (TXV). Three types of measurement sensors are used in this study which is tri-axial accelerometer for the vibration, tachometer for the engine rpm tracking and microphone for the noise measurement. Two types of operating conditions are taken into consideration, when engine running at 850 rpm (idle) and 850-3000 rpm (tracking) conditions and a constant blower speed is applied for both conditions. The result shows that, the hissing type of noise is determined at the frequency range of 4500-5000 Hz for the both idle and running conditions, whereby the vibration at the Evaporator Inlet is the most significant compared to the Evaporator Core and TVX components. The vibration of the Evaporator Inlet shows the drastic vibration increment between 1000-1500 rpm and getting worse towards 3000 rpm. This result is validated with the 3D colour of noise waterfall analysis, whereby the hissing noise shows the dominant result in the frequency range of 4500-5000 Hz.

Keywords: Hissing noise, HVAC system, vibration, rpm tracking and Evaporator Inlet.

INTRODUCTION

Noise and vibration of the heating and ventilating air conditional (HVAC) system is the common problem for most of the automotive manufacturers. There is a lot of noise and vibration types that can be produced from the HVAC system to the vehicle such as humming, hissing, clicking, air-rush and many more. These noises can contribute to the significant issue for driving comfort. The automotive air conditional system has several assembly parts such as compressor, condenser, radiator fan, drier, thermal expansion valve (TXV), evaporator and blower, as shown in Figure 1

(Autointehbox, 2015). All these parts can contribute to the specific noise and vibration problem for the vehicle.

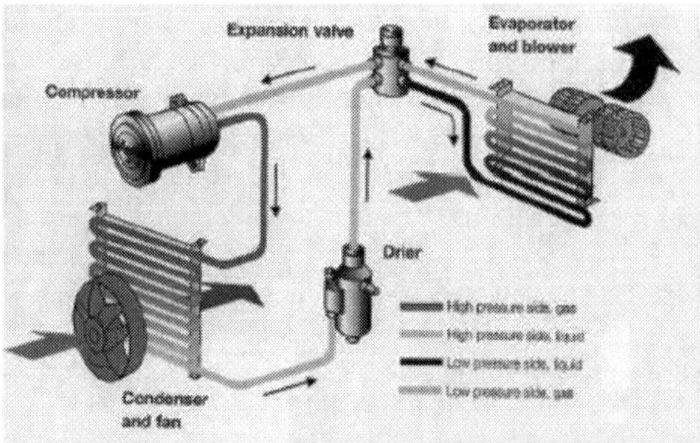

Figure 1. Common vehicle HVAC system (Autointehbox, 2015)

The current design of the HVAC system is based on the circulation of liquid refrigerant and water to perform the function of cooling and heating for the vehicle. There are two main sections of the HVAC system, one side dealing with the liquid such as condenser, drier and TXV while another side dealing with the air such as compressor, evaporator and blower. Most of the liquid side of the system is located within the engine location of the vehicle, with only two liquid-to-air heat exchangers positioned in the air handling unit inside the vehicle cabin. Both liquid and air sides can produce noise and vibration problem from various components (Kamiyama and Uomoto, 1997). One of the major HVAC components that contributed to the noise and vibration problem is a compressor. Since compressor is the main component that turned the low-pressure air to the high-pressure air, there is a high possibility in creating the noise such as humming and clicking noises. These noises are created during the compressor engagement and disengagement period of the overall HVAC system. This is the most common noise problem to the HVAC system which can damage the compressor part if no necessary action been taken (Mazlan, et al., 2018).

In order to determine the root cause and type of noise and vibration of the HVAC system, a proper set-up and methodology must be carried out. The study done by (Mavuri, et al., 2008) shows the proper methodology in measuring the vehicle HVAC cabin noise such as the testing parameters involved, instrumentation to be used and the set-up procedure. In other study, the lab scale rig has been set-up to determine the noise created from the air handling unit of the HVAC system. Based on results of this study, it was proven that by increasing the fan will increase the total noise level and magnitude of the predominant noise spectral peaks. Another finding from the study is regarding the cover of motor vent passage hole and the seal of weld gap which do not change the total noise level but do reduce the noise spectral peaks at the blade passing frequencies which also affected the cavity resonance peaks at the low frequencies (Wang and Watkins, 2007).

Since there are varies type of noise and vibration that can be induced from the HVAC components, this study only focused on one type of noise which is hissing-type noise that generated from one of vehicle HVAC components. The study consists of in-vehicle vibration and noise measurements at related HVAC components such as Evaporator Inlet, Evaporator Core and TXV components in order to determine the root cause of the hissing-type noise and vibration.

METHODOLOGY

Experimental Set-up

In this section, the experimental setup for the in-vehicle hissing type noise and vibration measurement are discussed. The detail of the test flow chart and test locations of the vehicle HVAC system for the vibration and noise measurement is shown in Figure 2 (a) and (b), respectively. In Figure 2 (a), there are three typed of sensors that been used which is accelerometer for the vibration (i.e., acceleration) measurement, microphone for the noise measurement and tachometer for the rpm measurement. The signal from these sensors are recorded using the DAQ (LMS SCADAS hardware) and then the data is analyze using LMS Test.Lab software to generate the vibration, noise and rpm related figures. The detail analysis using Sound Diagnosis workbench is performed with the frequency filtering function to extract the actual hissing-type noise and vibration characteristics. The locations of measured HVAC components for the vibration and noise are shown in Figure 2(b), where there are total of three locations for the HVAC system components such as Evaporator Inlet, Evaporator Core and TXV components. These locations are the main target of the vibration induced noise for the hissing-type noise of the vehicle HVAC system.

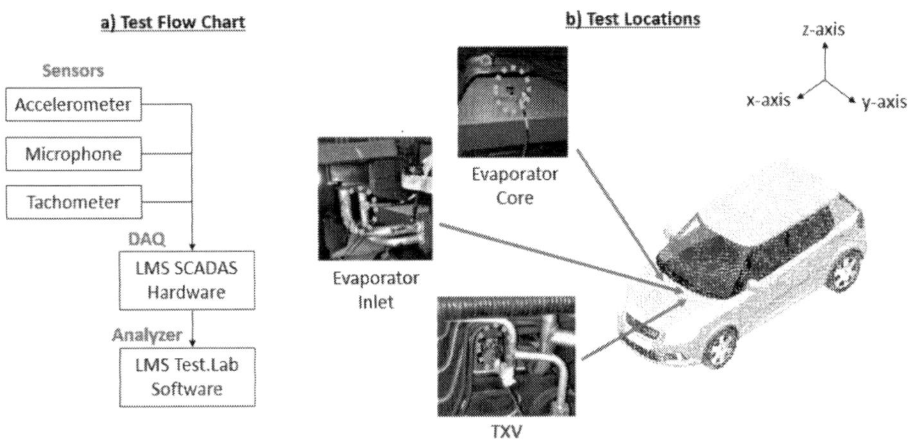

Figure 2. (a) Test flow chart and (b) Test locations of the vehicle HVAC system

Testing Conditions for Hissing-type Vibration and Noise Measurement

In order to determine the root cause of the hissing-type noise and vibration of the HVAC system, the testing conditions have to be properly decided. Table 1 shows the testing conditions for the vibration, noise and rpm tracking measurement (hissing-type) for the vehicle HVAC system. From the table, there are two test conditions applied, which is in idle (Engine speed at 850 rpm) and in-rpm tracking (Engine speed increasing from 850 to 300 rpm). For both conditions, engine status of "On", air conditional of "On" and blower speed of 1 are applied. Since this is a fundamental investigation to determine the root cause of the hissing-type noise, only one level of blower speed (i.e. 1/4 speed) is considered. Further investigation for varies blower speed setting can be performed after determining the root cause of the noise.

Table 1. Testing conditions of the HVAC system (Hissing type).

Type of noise	Test conditions	Engine status	Air conditional status	Blower speed	Locations
Hissing	Idle (850 rpm)	On	On	1	- Evaporator Inlet - Evaporator Core - TXV
	Rpm tracking (850-3000 rpm)	On	On	1	

RESULTS AND DISCUSSION

Hissing-type Noise and Vibration Analyses

In this section, the result of hissing-type noise and vibration in terms of sound pressure level (SPL) and vibration amplitudes are presented using frequency domain and 3D color waterfall analysis methods. Figure 3 shows the vibration amplitude (i.e., acceleration) against rpm tracking of the Evaporator Inlet component. As shown in the figure, the acceleration of the Evaporator Inlet is low at 1 m/s^2 during the idle condition of 850 rpm. When the engine is started to ramp from 850 – 3000 rpm, the vibration amplitude of vibration increases drastically from 1000 – 1500 rpm of range with the maximum amplitude of 1.8 m/s^2. After 1500 rpm, the vibration signal started to be fluctuated and getting worse for the range of 1500 – 2500 rpm as shown in Figure 3. From this result, the hissing-type noise is clearly heard from the subjective feeling during this range of rpm. By referred to the subjective feeling and vibrational response in Figure 3, the result is further analyzed using the frequency domain and 3D color waterfall analysis methods.

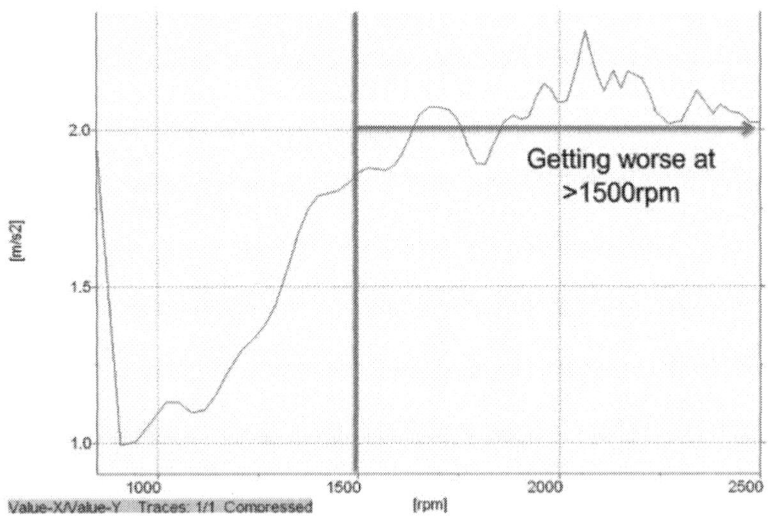

Figure 3. Vibration against rpm tracking of the Evaporator Inlet

Figure 4 (a) shows the 3D waterfall graph of the frequency-rpm-SPL of the microphone mounted at the driver left ear inside the vehicle cabin. From the figure, it is clearly shows that, the SPL is dominant at the frequency range of the 4500 – 5500 Hz when the engine rpm is tracking in the range of 1000 – 2000 rpm. The highest value of SPL is recorded at 0.05 x 10^{-3} Pa for the range of red color mapping. The result is further verified with the vibration amplitude comparison between Evaporator Inlet, Evaporator Core and TXV components, as shown in Figure 4 (b). From the figure, it is clearly shows that, the vibration amplitude for all three HVAC components are significant at the frequency range of 4500 – 5500 Hz with the highest vibration value produced by the Evaporator Inlet component in *z*-axis direction with 0.18 m/s^2 of acceleration. This frequency range, if compared in term of subjective hearing perspective is clearly indicates the hissing-type of noise of the HVAC system components.

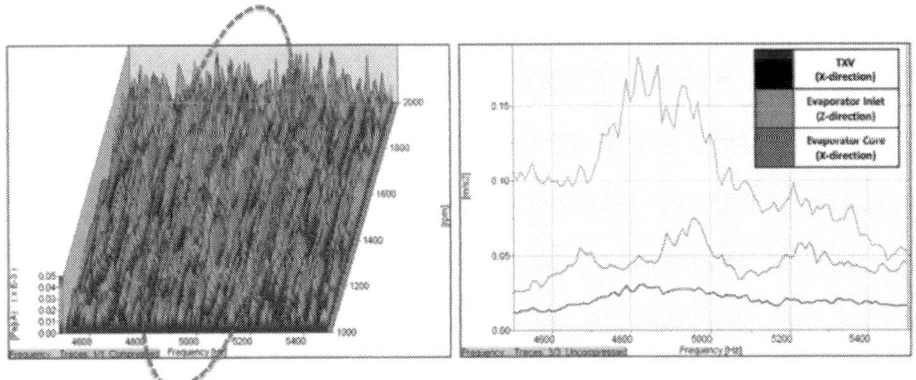

a) Noise measurement **b) Components vibration**
(Microphone at driver left ear)

Figure 4. (a) SPL waterfall noise and (b) FFT of vibration for three HVAC components

CONCLUSION

In this study, the main objective to determine the root cause of the hissing-type noise and vibration of the vehicle HVAC components has been achieved. Table 2 shows the summary of the result, where the Evaporator Inlet has contributed a significant vibration amplitude of 0.18 m/s^2 in the frequency range of 4500 – 5500 Hz for rpm tracking test condition. This result is verified using 3D color waterfall analysis and the SPL level shows a significant noise value during these frequency and rpm tracking ranges. In term of subjective hearing perspective, this frequency range has clearly indicated the hissing-type noise of the HVAC system components.

Table 2. Result summary of the HVAC components (Hissing-type).

Type of noise	**Test conditions**	**Noise and vibration frequency ranges**	**Highest noise contribution parts**
Hissing	Idle (850 rpm)	4500 – 5500 Hz	Evaporator Inlet
	Rpm tracking (850-1400 rpm)		

ACKNOWLEDGEMENTS

The authors would like to acknowledge Universiti Sains Malaysia and Kementerian Pengajian Tinggi for providing the financial assistance under Short-term and FRGS grants (304/PMEKANIK/60313052 and 203/PMEKANIK/6071370) and Proton Holdings Berhad for their laboratory facilities and technical support for this project.

REFERENCES

Autointehbox.com (2015). What you don't know about your automotive air conditioning system. [Accessed on January 2017]. Available from World Wide Web: *http://blog.autointhebox.com/what-you-dont-know-about-your-automotive-air-conditioning-system.html.*

Kamiyama, N. and Uomoto, M. (1997). Development of low noise blower fan saeki, *SAE paper*, No 971842.

Mazlan, A. Z. A, Hamdan, M. H., Md. Isa, M. S., Man, S., Paiman, M. A. R. and Abd. Ghapar, M. Z. (2018). Characterization of the humming-type noise and vibration of the automotive HVAC system. *Proceedings of the 6th International Conference on Noise, Vibration and Comfort, Malaysia.*

Mavuri, S. P., Watkins, S., Wang, Xu., Hill, S. S. and Weymouth, D. (2008). An investigation and bench marking of vehicle HVAC cabin noise. *Proceedings of 2008 SAE World Congress, U.S.*

Wang, X. and Watkins, S. (2007). Noise refinement solutions for vehicle HVAC, *SAE paper*, 07NVC-112.

DEEP NEURAL NETWORK TOOL CHATTER MODEL FOR ALUMINUM SURFACE MILLING USING ACOUSTIC EMMISION SENSOR

M. Abul Hasan[1], Muhamad-Husaini Abu-Bakar[1*], Rizal Razuwan[1], Zainal Nazri[1]

[1]System Engineering and Energy Laboratory,
Universiti Kuala Lumpur Malaysian Spanish Institute,
*Email: muhamadhusaini@unikl.edu.my
Phone: +6044035199; Fax: +6044035201

ABSTRACT

Chatter is a self-excited vibration in any machining processes which contributes to the system instability due to resonance and resulting in an inaccuracy in machining product. Due to demand for a high precision product, industries are nowadays moving towards implementing a tool monitoring system as a feedback. Currently, an electromagnetic sensor was used to detect chatter in tools, but this sensor introduces a drawback such as bulky in size, sensitive to noise and not suitable to be implemented in the small machining center. This paper aims to propose a chatter identification model for face milling tool based on acoustic emission data for tool monitoring system. Acoustic emission data is collected at four level of cutting depth in milling with linear tool path movement on aluminum T6 6061 materials. The Deep Neural Network (DNN) model was developed using multiple deep-learning frameworks for the chatter detection system. This model approach shows a good agreement with experimental data with 4% error. As a conclusion, the DNN chatter identification model was successfully developed for the aluminum milling process applications. This finding is essential for anomaly detection during machining process and able to suggest for a better machining parameter for the aluminum machining process.

Keywords: Tool Chatter, Deep Neural Network, Acoustic Emission, Anomaly detection.

INTRODUCTION

Chatter identification has been available in many countries especially in the first-class country. There are many applications of chatter identification which have been studied and developed by many researchers such as detect breakages, detect vibration, and monitoring system. In addition, chatter identification also applicable in machine vision, tool toughness, dimensional accuracy, and surface roughness application. Chatter is a self-excited vibration in any machining processes which contribute to the system instability due to resonance and resulting in any damage to the workpiece also machine tool [1]. Thus, the machining process today's need the chatter identification system.

Nowadays, manufacturing enterprises are moving in line with Industry 4.0 which turning towards full automation, reliable, online, and robust monitoring system for more productivity, quality, precision, and reduced costs [2]. In practice, tool monitoring system utilizes a chatter identification result to determine a degree of vibration in the machining process. Tool monitoring system is the system which manages, monitor and analyzes the process of machining process for detecting the defect in the process or tools [3]. Indeed, in maintaining high quality in automated and un-manned manufacturing, the tool integrity must be maintained, tool condition monitoring is required [4].

Chatter identification was known as fundamental techniques in tool monitoring system. Many types of sensor which emitted vibration input data such as an accelerometer, electromagnetic and ultrasonic sensors. Numbers of researcher suggest using an electromagnetic sensor to detect chatter in a tool [5], but this kind of sensors introduce a drawback such as bulky in size, sensitive to noise and not suitable to implement in a small machining center.

In that case, the spectral domain obtained from Fourier analysis was the perfect alternative used to extract frequency-domain features of measured vibration signals from acoustic emission sensors [6]. In the last decade, NNs have been applied to monitor machining processing, including turning, milling and grinding machines [7]. Different from traditional methods, deep learning is a new machine learning methods, which has the great capacity to automatically learn the valuable features and inference from the data [8].

Hence, the aim of this research is to develop a chatter identification model for tool monitoring system using Deep Neural Network (DNN) model which using acoustic emission input data. The proposed model was predicted to gives an accurate solution of chatter identification system. This project goal was to develop an experimental setup for proposed model in the milling process which divided into several phases.

METHODOLOGY

Experimental Setup

Experimental setup of this project has been designed and set-up for tool monitoring system. However, identification of component parameters needs verification before proceeding to the experimental setup. Firstly, the elements in experimental setups must be accurate and precise for monitoring the chatter. Therefore, the calibrations process must be conducted before further experimenting to make sure the components are in good condition for monitoring the tools wear and specimen surface roughness.

Table 1. The component of experimental

No	Component	Type
1	CNC Milling Machine	Fagor C355
2	Flat End Mill	16mm 2 Flutes (HSS)
3	Aluminium Block	100 mm x 100 mm x 25 mm
4	Computer Monitor	HP EliteDisplay E201
5	Preamplifier	Mistras Preamplifier
6	AE System	Mistras Micro-ii Digital
7	AE Sensor	WD - 100-900 KHZ Wideband Differential AE Sensor

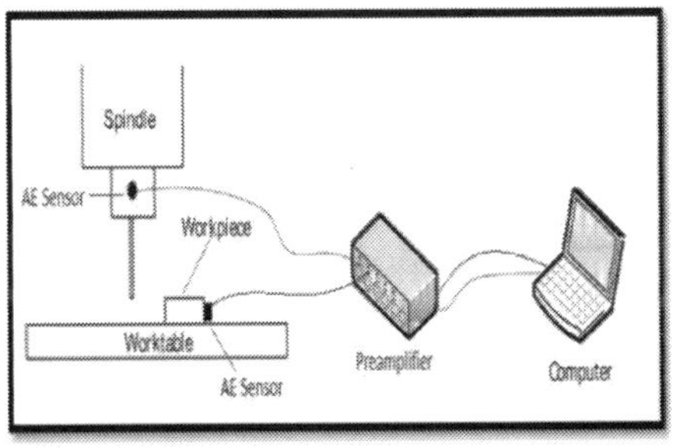

(a)

(b)

Figure 1. (a) Schematic diagram of experimental setup (b) Picture of experimental setup

Next, the milling tests were performed with a Fagor C355 CNC milling machine as shown in Figure 1 (a) and (b). The specimen of aluminum T6 6061 material within a specific size was clamped on the work table and High Speed Steel (HSS) end mill cutter with two flutes and diameter 16mm was used. The measurements were conducted using Mistras Micro-ii Digital system combined with its software, AEWin. Acoustic emission sensors were mounted on the spindle housing to measure the vibration signals during the milling process on the work table which the specimen was clamped to record the vibration signals during the machining processes. The integrated acoustic emission sensor process and receive the vibration signals and convert it's into data.

Since the occurrence of chatters was directly correlated with the depth of the cutting, each experiment have been conducted at a constant spindle speed of 2000 r/min, a feed rate of 747 mm/min, and with a different depth of the cutting at 0mm, 1.0 mm, 3.0mm and 5.0 mm, respectively. All the experiments were slotting without cutting fluids. The parameters for chatter identification was classified into steady, chatter level 1, chatter level 2 and chatter level 3 as shown in Table 2.

Table 2. The parameters for chatter identification system

Cutting Condition	Depth of a cut (mm)	Spindle Speed (rpm)	Feedrate (mm/min)	Remarks
Steady State	0	2000	747	Without coolant fluid
Chatter Level 1	1	2000)	747	Without coolant fluid
Chatter Level 2	3	2000	747	Without coolant fluid
Chatter Level 3	5	2000	747	Without coolant fluid

Data Analysis

The signal data that has been emitted by AE sensors was transmitted through preamplifier for making a signal fit for the next experimental stage in the signal chains and performing a filtering process for an unwanted signal. This approach was performed due to a noise in the system and was overcome by adding a preamplifier for minimizing a noise in the system compared to raw signals. The tuned signals have been transmitted to the computer database for the data analysis procedures.

The acquisition of the data between steady and chatter condition that was acquired has been transformed and prepared for time-frequency data. The data preparation, transformation, and visualization were produced using multiple open-source tools which specialized in machine learning platforms such as Rstudio desktop version '1.0.153' [9], SAS University Edition Studio, Orange [10], and Spyder Python integrated development environment (IDE).Furthermore, the proposed model for chatter identification was also developed using tool mentions before and the model was trained and evaluated with the experimental data.

Chatter Identification Model

The tools used for developing the model was a GPU-accelerated and thus, reducing the time need for computing the developed model. The requirements for using multiples deep-learning frameworks such as Keras [11], TensorFlow [12], and Microsoft Cognitive Toolkit (CNTK) [13] was installing all dependencies and set-up our model environment. The advantages of using multiple deep-learning frameworks were less computing time, customized model, a high-level abstraction layer, and fast prototyping. The DNN model was developed by using Long Short Term Memory (LSTM) network from Keras model which can outperform the state-of-the-art anomaly detection algorithms on time-series data. Figure 2 shows the basic LSTM network illustration.

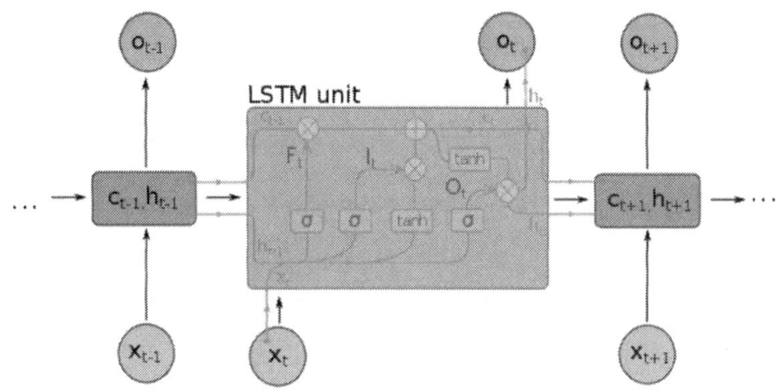

Figure 2. Basic example of LSTM network

RESULTS AND DISCUSSION

Analysis of Chatter Condition

Figure 3 (a) and (b) shows the visualization and analysis of chatter condition from the first stage into the second stage using chatter identification system that was developed. The process of identification was implemented during machining process on CNC milling machine. The result shows the frequency of specimen cutting condition against time taken for different stages.

As shown below for specimen condition in Figure 3(a), there was an arising chatter that was caught by the chatter identification system which starts from steady state and suddenly arose to a peak point during the machining process. Based on this result, this shows that the chatter can be identified using the developed model of chatter identification system. Moreover, in Figure 3(b) the pattern of chatter condition was different than Figure 3(a) which was obviously showing the differences in amplitude of the chatter frequency. This was caused by the difference in cutting depth which clearly proved that the increase of cutting depth influence the peak of anomaly chatter and tool wear rate.

In the middle of chatter pattern in Figure 3(b), showed that the amplitude starts to rise back after a colossal drop and then suddenly drop before begin to increase in the late time of the second stage. These anomalies were identified and visualized using the developed chatter identification system for tool monitoring system which can be used for the system feedback to ensure high precision product.

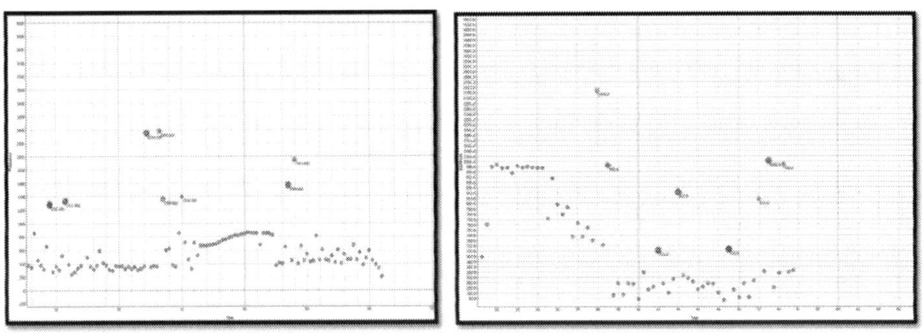

(a) (b)

Figure 3. (a) First stage of chatter condition (b) Second stage of chatter condition

Analysis of Steady Condition

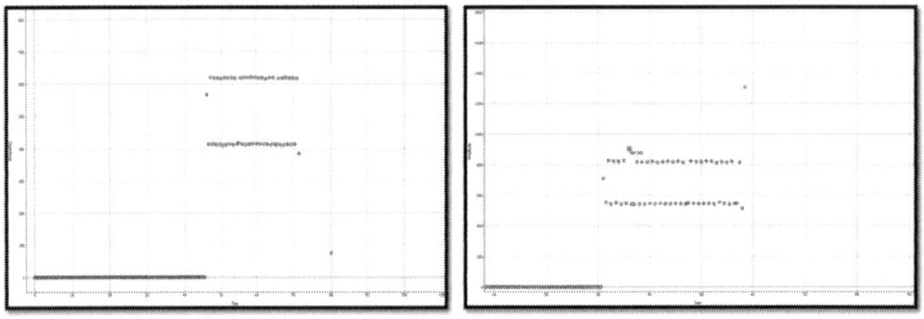

(a) (b)

Figure 4. (a) First stage of steady condition (b) Second stage of steady condition

Based on these result in Figure 4(a) and (b), there were only minor anomalies detected during steady condition which caused less tool wear condition during the machining process. Figure 3 (a) shows no arising chatter identified by the chatter identification system and reveal that the machining process was in steady state and minor anomalies. The maximum frequency detected in the first stage was 876 Hz which is less than the second stage maximum frequency, 904 Hz.

Comparison of chatter and steady conditions

From the previous study, it was explained that the most frequency components directly in the range of 500 Hz to 3000 Hz [14]. Then, the distribution of the frequency starts to concentrate on a range of chatter frequencies condition. Based on the result in Figure 5, the most of the frequencies have been detected in the range 720-2478 Hz. Nonetheless, the distribution of frequency components have been not concentrated in range the chatter frequencies, but it was freely chattering identification regarding after tool was passing through and contacted on the specimen. The proposed method used with Orange [10] was not precisely same output as Fourier spectra but this method significantly improves the accuracy of tool monitoring system by gaining 12.4% from previous research.

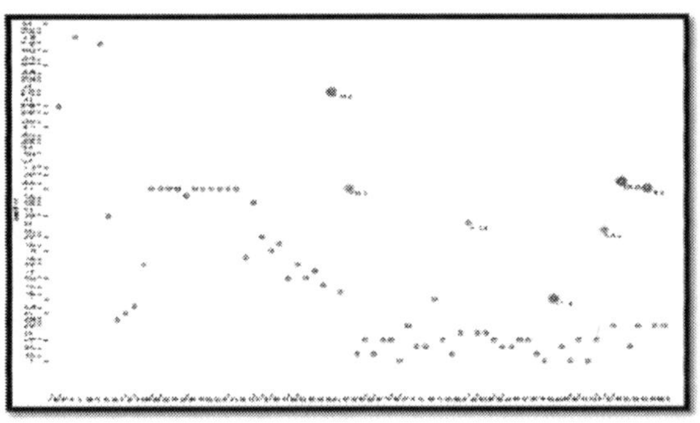

Figure 5. Chatter identification from experimental data

CONCLUSION

Based on the resuts, the proposed model of chatter identification system which was developed with DNN model was able to detect the chatter in the aluminum surface milling process using AE sensors. Besides that, this model was proved in this experimental more reliable and robust for more productivity, quality, precision, and eventually reducing the cost. This model also able to define the symptom and phenomenon of chatter occurrence in both time and frequency domain. Lastly, the advent of Artificial Intelligence (AI) proves that it can solve a complex problem and can be used for better human futures.

REFERENCES

[1] Wan, S., Li, X., Chen, W., & Hong, J. (2018). Investigation on milling chatter identification at early stage with variance ratio and Hilbert–Huang transform. *International Journal of Advanced Manufacturing Technology*, *95*(9–12), 3563–3573. https://doi.org/10.1007/s00170-017-1410-y

[2] Wan, S., Li, X., Chen, W., & Hong, J. (2018). Investigation on milling chatter identification at early stage with variance ratio and Hilbert–Huang transform. *International Journal of Advanced Manufacturing Technology*, *95*(9–12), 3563–3573. https://doi.org/10.1007/s00170-017-1410-y

[3] Abellan Nebot J and Romero Subiro´ n F (2010) A review of machining monitoring systems based on artificial intelli- gence process models. The International Journal of Advanced Manufacturing Technology 47: 237–257.

[4]Ratava, J., Lohtander, M., & Varis, J. (2017). Tool condition monitoring in interrupted cutting with acceleration sensors. Robotics and Computer-Integrated Manufacturing, 47(November), 70–75. https://doi.org/10.1016/j.rcim.2016.11.008

[5] Ahrens, M., Dagen, M., Denkena, B., & Ortmaier, T. (2016). An Active Damping Method for Chatter Vibration in Plunge Grinding Using Electromagnetic Actuators. Procedia CIRP, 46, 197–200. https://doi.org/10.1016/j.procir.2016.03.200

[6] Samanta B, Al-Balushi KR and Al-Araimi SA (2006) Artificial neural networks and genetic algorithm for bearing fault detection. Soft Computing – A Fusion of Foundations, Methodologies and Applications 10: 264–271.

[7] Yao Z, Mei D and Chen Z (2010) On-line chatter detection and identification based on wavelet and support vector machine. Journal of Materials Processing Technology 210: 713–719.

[8] Wan, S., Li, X., Chen, W., & Hong, J. (2018). Investigation on milling chatter identification at early stage with variance ratio and Hilbert–Huang

transform.*International Journal of Advanced Manufacturing Technology*, *95*(9–12), 3563–3573. https://doi.org/10.1007/s00170-017-1410-y

[9] RStudio Team (2015). RStudio: Integrated Development for R. RStudio, Inc., Boston, MA URL http://www.rstudio.com/

[10] Demsar J, Curk T, Erjavec A, Gorup C, Hocevar T, Milutinovic M, Mozina M, Polajnar M, Toplak M, Staric A, Stajdohar M, Umek L, Zagar L, Zbontar J, Zitnik M, Zupan B (2013) Orange: Data Mining Toolbox in Python. Journal of Machine Learning Research 14(Aug):2349−2353.

[11] Chollet, Franc¸ois. Keras. https://github.com/fchollet/keras, 2015.

[12] TensorFlow: Large-scale machine learning on heterogeneous systems, 2015, Software available from tensorflow.org.

[13] Frank Seide and Amit Agarwal. 2016. CNTK: Microsoft's Open-Source Deep-Learning Toolkit. In Proceedings of the 22nd ACM SIGKDD International Conference on Knowledge Discovery and Data Mining (KDD '16). ACM, New York, NY, USA, 2135-2135. DOI: https://doi.org/10.1145/2939672.2945397

[14] Cao, H., Zhou, K., & Chen, X. (2015). Chatter identification in end milling process based on EEMD and nonlinear dimensionless indicators. *International Journal of Machine Tools and Manufacture*, *92*, 52–59.

A LOW COST POTENTIOSTAT DEVICE FOR MONITORING AQUEOUS SOLUTION

S. N. H. Umar[1*], E. A. Bakar[1], N. M. Kamaruddin[1] and N. Uchiyama[2]

[1]School of Aerospace Engineering,
Universiti Sains Malaysia,
*Email: snhanisahumar@gmail.com
[2]Department of Mechanical Engineering,
Toyohashi University of Technology,
Aichi, Japan.

ABSTRACT

This study developed a new design of a low cost potentiostat circuit device. This device is an alternative electrochemical instrument applied for monitoring heavy metal ion in environmental water. It was developed to alleviate the cost burden of equipment procurement and due to the requirement for in-situ application since the existing commercialize devices are bulky and expensive. The main component of the device consist of electronics configuration of operational amplifier. The device was first modelled and simulated to acquire the design parameters and performance. The potential control algorithm was developed on open-source microcontroller platform. A dummy cell was used to validate the capabilities of the device.

Keywords: Potentiostat, Operational amplifier, Electrochemical, Electronics configuration.

INTRODUCTION

Water pollution is a serious threat to global health [1]. Heavy metal ions (HMI) especially lead (Pb), cadmium (Cd), mercury (Hg), chromium (Cr) and arsenic (As) are highly toxic even in small doses [2]. Industrial activities are the main source of HMI [3]. These HMI are not biodegradable and have a tendency to accumulate in living organisms, causing various diseases and disorders to the nervous, immune, reproductive and gastrointestinal systems [2]. Thus, detection of HMI and determining their quantities is a paramount importance.

Conventional HMI detection technique such as spectroscopic is accurate, but costly, bulky, time consuming and require a well-trained analyst [4], [5]. An electrochemical technique is a reliable alternative to spectroscopic and has gained more popularity as it is more economic, user-friendly, allow a simple procedures and suitable to fabricated on small circuit in the portable form for in-field applications [2].

A potentiostat is a crucial component as the electrochemical analysers. The main functions are to control the applied potential voltage for HMI excitation in solution and to measure the resulting current due to movement of electron in the solution [6]. For potential control process, three electrode system was used as the sensor electrodes; working electrode (WE), reference electrode (RE) and counter electrode (CE) [7]. The WE can be defined as the electrode whose potential is a function of the solution's concentration [8]. The RE maintain a constant voltage referred to the potential of the hydrogen electrode. Thus, no current is allowed to pass the RE. The CE served to complete the electric circuit and provided the information of the ion in term of current reading.

In order to analyse and process the information from sensor electrodes, potentiostat is used. Due to the available commercialize potentiostat are expensive, bulky and often exceed the practical needs, researchers tends to build its own using existing information [9]–[16]. Thus, the purpose of this study is to describe the development of a potentiostat device for the purposed of HMI detection.

METHODOLOGY

Concept of Main Potentiostat

The proposed potentiostat was designed for three electrode system which consist of electronics configuration mainly the operational amplifier and resistor. The potentiostat can be categories into six main components as shown in Figure 1. The potentiostat was first designed and simulate on MATLAB Simscape for design parameters and performance test.

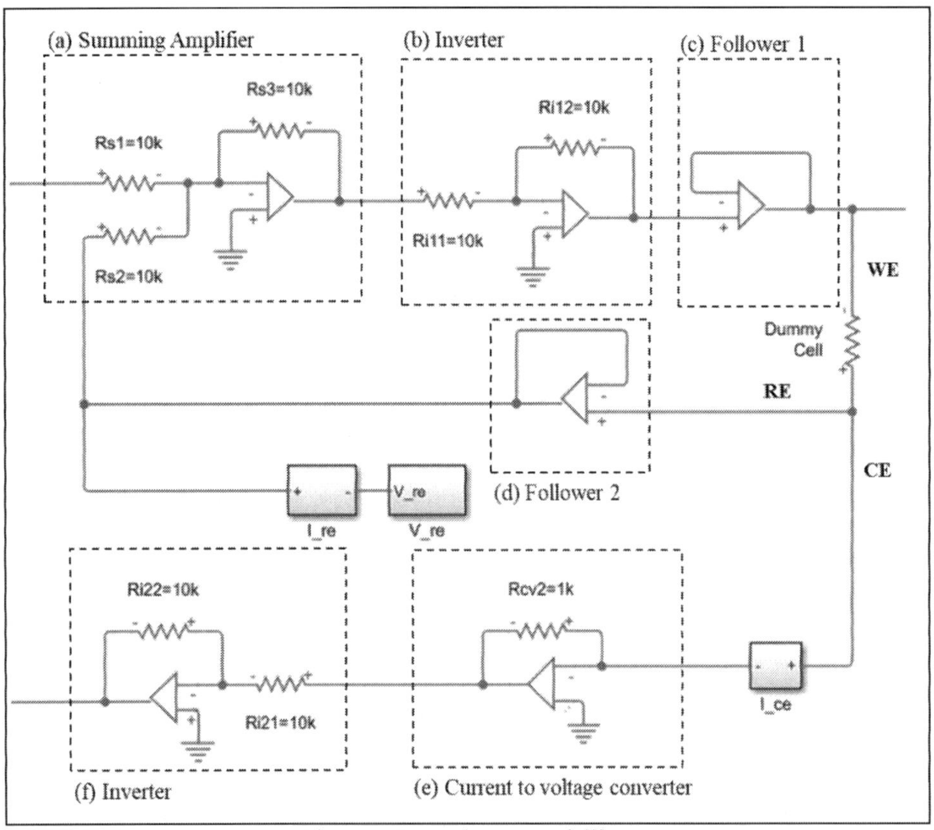

Figure 1 Potentiostat modelling.

Firstly, the applied potential voltage V_{in_1} passed through the summing amplifier. The main function of summing amplifier as in Figure 1(a), is to consider and provide the potential voltage, V_{in_2} sensed by RE as a feedback signal in potentiostat circuit. This is to ensure a sufficient excitation voltage supply at the WE due to the applied voltage at WE is always with respect to the RE. The general formulation for voltage summing is as below,

$$V_{out} = -\left(\frac{R_{s3}}{R_{s1}}V_{in_1} + \frac{R_{s3}}{R_{s_2}}V_{in_2}\right) \tag{1}$$

Since $R_{s_1} = R_{s_2} = R_{s_3} = 10k\Omega$, the output of summing amplifier can be written as,

$$V_{out} = V_{in_1} + V_{in_2} \tag{2}$$

The output produced by voltage summing is in inverted polarity. Thus, an inverter as in Figure 1(b) is used to rectify the polarity. The general formulation of the inverter,

$$V_{out} = -\frac{R_{i_1}}{R_{i_2}}V_{in} \tag{3}$$

Since $R_{i_1} = R_{i_2} = 10k\Omega$, the output of polarity inverter can be simplified as,

$$V_{out} = -V_{in} \tag{4}$$

A voltage follower as in Figure 1(c) and (d) provided a high input impedance to prevent significant current drawn that may lead to inaccurate potential voltage reading. The formulation of voltage follower can be written as,

$$V_{out} = V_{in} \tag{5}$$

The current signal from CE is measured using the transimpedance amplifier or current to voltage converter as in Figure 1(e). The formulation current to voltage converter is shown in Equation (6). The value of R_{cv2} determined the sensitivity range of the potentiostat. The inverter amplifier Figure 1(f) is cascade with transimpedance amplifier to rectify the output voltage polarity.

$$V_{out} = iR_{cv2} \tag{6}$$

Potentiostat Development

The designed potentiostat was implemented on a breadboard using LM324N Operational Amplifier. The circuit arrangement is shown in Figure 2 consist of three parts, (a) an ATmega328P Arduino Uno microcontroller for signal generator, parameter control and data acquisition. The second part is (b) a digital and analog converter (DAC) including unipolar to bipolar converter (UBC) and bipolar to unipolar converter (BUC). The third part (c) is the main potentiostat circuit as described in Figure 1.

For hardware implementation, the DAC part was applied as a connector between the Arduino Microcontroller and the main potentiostat part. The detail description of DAC is shown in Figure 3. A low pass filter is used to convert the square wave signal generated by microcontroller pulse width modulation (PWM) into an analog voltage. The relation between PWM and desired analog voltage, V_{in} are stated in Equation (7).

$$V_{in} = high\ voltage\ of\ PWM\ signal\ \times duty\ cycle \quad (7)$$

With the cut off frequency, f_c of,

$$f_c = \frac{1}{2\pi RC} \quad (8)$$

The PWM voltage of Arduino Uno board can only provide DAC voltage range from 0V to 5V. Since electrochemical experiment usually required a potential range from -1V to +1V, UBC is used as voltage mapping as shown in **Error! Reference source not found.** and the equation can be written as below,

$$V_{out} = V_{in}\left(\frac{R_2}{R_1 + R_2}\right)\left(1 + \frac{R_4}{R_4}\right) - V_2\frac{R_4}{R_3} \quad (9)$$

Substituted $R_1 = 20k\Omega, R_2 = R_3 = 10k\Omega$, and $R_4 = 2k\Omega$, Equation (9) is simplified as,

$$V_{out} = 0.4V_{in} - 1 \quad (10)$$

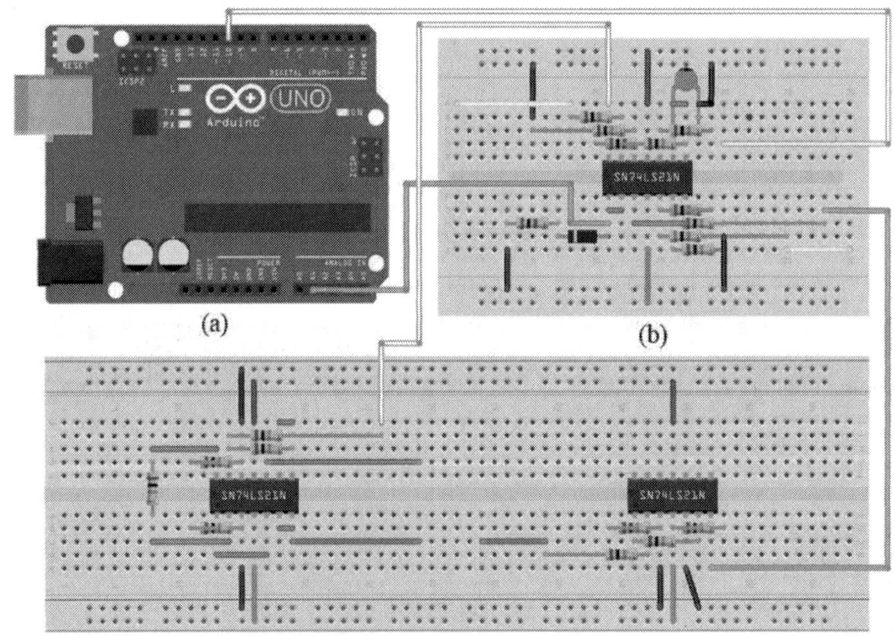

Figure 2 Potentiostat development board (a) microcontroller, (b) digital and analog converter and (c) main potentiostat circuit.

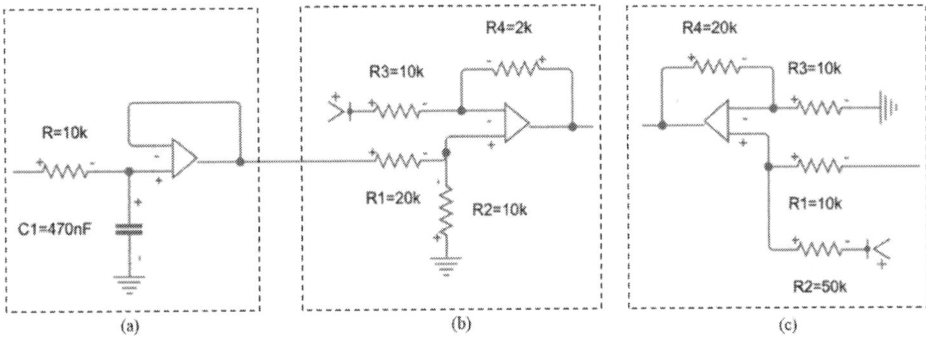

Figure 3 Components of digital and analog converter (a) low pass filter, (b) unipolar to bipolar converter and (c) bipolar to unipolar converter.

In Figure 4 shows the output of low pass filter indicated as Vin and the output of UBC indicated as Vout.

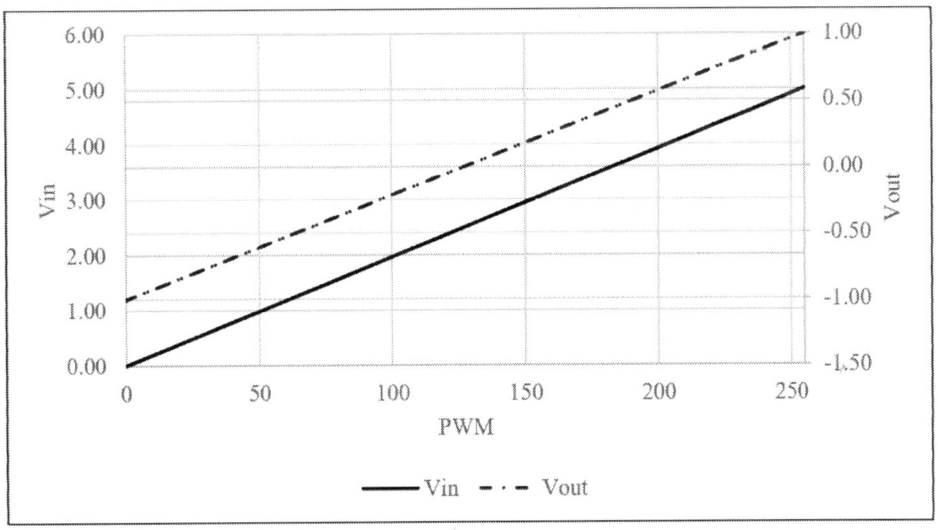

Figure 4 DAC and UBC output.

In order for microcontroller to analyse and process the signal from sensor electrodes, BUC as shown in Figure 3(c) is used to convert the voltage signal range (-1V to +1V) from main potentiostat to microcontroller range (0V to +5V). The formulation of the converter is shown below,

$$V_{out} = \left(V_{in}\frac{R_2}{R_1 + R_2} + V_2\frac{R_1}{R_1 + R_2}\right)\left(1 + \frac{R_4}{R_3}\right) \tag{11}$$

Substitute resistance value, rewrite Equation (11) as,

$$V_{out} = 2.5(V_{in} + 1) \tag{12}$$

https://doi.org/10.1051/matecconf/201821704001

RESULTS AND DISCUSSION

The capability of the potentiostat device was verified using a dummy cell. The dummy cell is a device replicated the dynamics of electro-analytical cells using passive electrical components such as resistors and capacitors in certain configurations. Linear sweep voltammetry was applied to evaluate the performance of the potentiostat device. Dummy cell of 1kΩ, 2kΩ and 10kΩ were connected between WE and RE and CE.

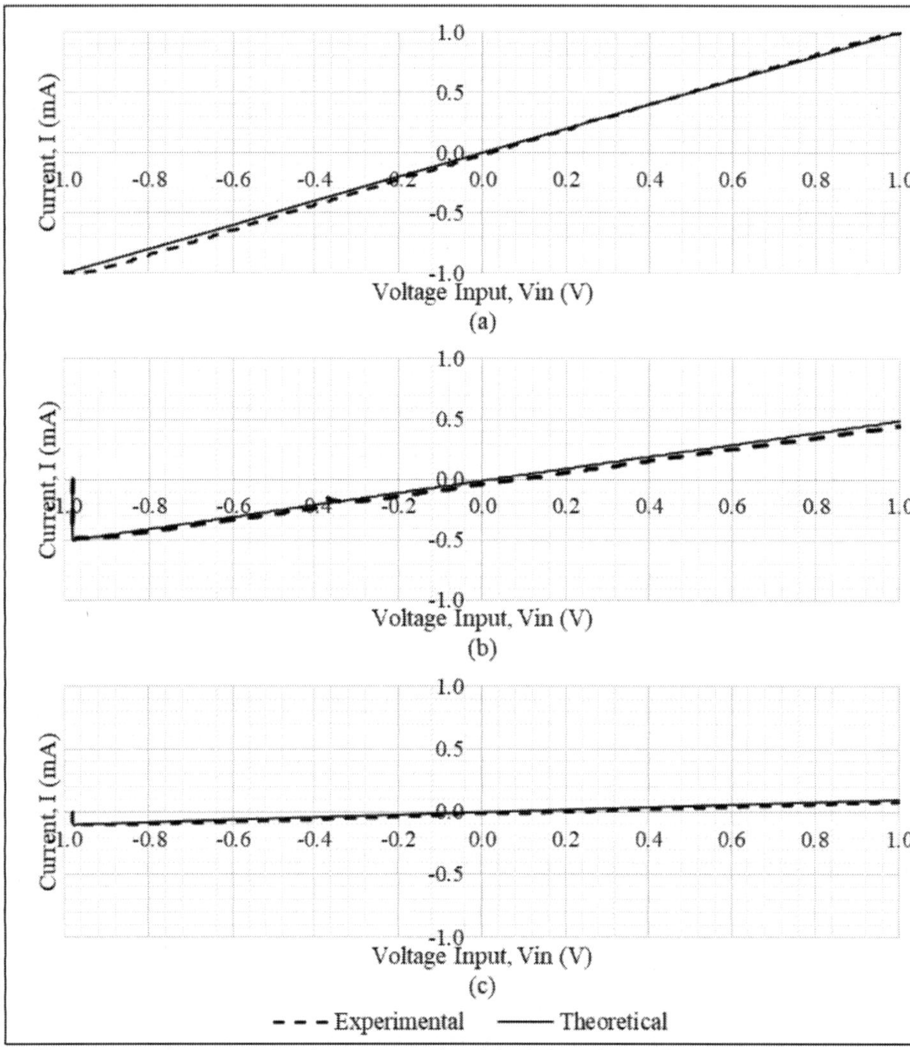

Figure 5 Resistor polarization curve using (a) 1kΩ resistor, (b) 2kΩ resistor and (c) 10kΩ resistor.

When a potential is swapped between the leads of a resistor, the measured current can be calculated based on the apply voltage input, V_{in} at the WE and is expected to show a linear response. As seen in Figure 5, the expected behaviour is observed. A similar trend was seen between experimental and theoretical. The polarization resistance was obtained as the inverse value of slop, $m = 1/R_p$ with $R_p = 0.962k\Omega, 2.099k\Omega$ and $10.121k\Omega$ for dummy cell 1kΩ, 2kΩ and 10kΩ respectively. These values were in a good agreement with the 10% tolerance resistor used. The mean error between experiment and theoretical is small up to 1.30E-05 as shown in Table 1.

Table 1 Mean error of validation test.

Dummy validation test value, R	Voltage input, Vin	Mean error, e
$R = 1k\Omega$	from -1V to +1V	2.41E-05
$R = 2k\Omega$	from -1V to +1V	2.95E-05
$R = 10k\Omega$	from -1V to +1V	1.30E-05

CONCLUSION

In this study, the low cost (< RM100) potentiostat device was developed using ATmega328P Arduino Uno microcontroller, LM324N operational amplifier, resistors and capacitor. Based on verification test, the designed potentiostat was capable to work as desire with similar trend and small mean error (up to 1.30E-05) between theoretical and experimental.

For further development study, it is highly recommend to improve in the following areas: (1) perform device validation using an actual electrochemical setup, (2) construction of the device using a custom etched printed circuit board and (3) higher portability for use in low-resource settings for researches and potential point-of-use care analysis.

ACKNOWLEDGEMENTS

The authors would like to acknowledge Universiti Sains Malaysia (USM) and Ministry of Education (MOE) for providing the financial assistance under RU Top-Down research grant (1001/PAERO/6740041) and awarding her USM Academic Staff Training Scheme (ASTS).

REFERENCES

[1] P. Wu, G. Vazquez, N. Mikstas, S. Krishnan, and U. Kim, "Aquasift: A low-cost, hand-held potentiostat for point-of-use electrochemical detection of contaminants in drinking water," *2017 IEEE Glob. Humanit. Technol. Conf.*, pp. 1–4, 2017.

[2] B. K. Bansod, T. Kumar, R. Thakur, S. Rana, and I. Singh, "A review on various electrochemical techniques for heavy metal ions detection with different sensing platforms," *Biosens. Bioelectron.*, vol. 94, no. January, pp. 443–455, 2017.

[3] F. Fu and Q. Wang, "Removal of heavy metal ions from wastewaters: A review," *J. Environ. Manage.*, vol. 92, no. 3, pp. 407–418, 2011.

[4] M. Nizar *et al.*, "An Electroanalytical Instrument Equipped with Wireless Communication Network and Graphical User Interface for Real-time Monitoring Wastewater Status from Batik Industry," *J. Mech. Eng. Vo l SI*, vol. 4, no. 2, pp. 153–170, 2017.

[5] M. B. Gumpu, S. Sethuraman, U. M. Krishnan, and J. B. B. Rayappan, "A review on detection of heavy metal ions in water - An electrochemical approach," *Sensors Actuators, B Chem.*, vol. 213, no. 2015, pp. 515–533, 2015.

[6] Ametek Scientific Instrument, "A Review of Techniques for Electrochemical

Analysis," *Princeton Applied Research*. [Online]. Available: http://nanobioelectronics.weebly.com/uploads/3/8/8/6/38862535/review_of_techniques_for_electrochemical_analysis.pdf.

[7] T. Arevalo-Ramirez, C. C. Torres, A. C. Rosero, and P. Espinoza-Montero, "Low cost potentiostat: Criteria and considerations for its design and construction," *Proc. 2016 IEEE ANDESCON, ANDESCON 2016*, 2017.

[8] D. Harvey, *Modern Analytical Chemistry*. 2000.

[9] G. N. Meloni, "Building a microcontroller based potentiostat: A inexpensive and versatile platform for teaching electrochemistry and instrumentation," *J. Chem. Educ.*, vol. 93, no. 7, pp. 1320–1322, 2016.

[10] A. Ainla, M. P. S. Mousavi, M. Tsaloglou, J. Redston, G. Bell, and M. T. Ferna, "Open-Source Potentiostat for Wireless Electrochemical Detection with Smartphones," *Anal. Chem.*, 2018.

[11] C. Y. Huang, H. T. Huang, and R. T. Yuan, "Design of a portable mini potentiostat for electrochemical biosensors," *Proc. 2017 IEEE 2nd Adv. Inf. Technol. Electron. Autom. Control Conf. IAEAC 2017*, pp. 200–203, 2017.

[12] D. L. Agulto, J. C. A. Guinto, L. K. G. Palomo, J. F. Villaverde, and M. A. C. Fabro, "Design of Microcontroller-based Potentiostat with Performance Validation using Auto Lab Eco Chemie Dummy Cell 2."

[13] A. A. Rowe *et al.*, "Cheapstat: An open-source, 'do-it-yourself' potentiostat for analytical and educational applications," *PLoS One*, vol. 6, no. 9, p. 7, 2011.

[14] M. D. M. M. Dryden and A. R. Wheeler, "DStat: A Versatile, Open-Source Potentiostat for Electroanalysis and Integration," *PLoS One*, vol. 10, no. 10, p. e0140349, Oct. 2015.

[15] K. Christidis, K. Gow, P. Robertson, and P. Pollard, "Intelligent potentiostat for identification of heavy metals in situ," *Rev. Sci. Instrum.*, vol. 77, no. 1, pp. 1–4, 2006.

[16] P. Bezuidenhout, S. Smith, K. Land, and T. H. Joubert, "A low-cost potentiostat for point-of-need diagnostics," in *2017 IEEE AFRICON: Science, Technology and Innovation for Africa, AFRICON 2017*, 2017, pp. 83–87.

MOTION ANALYSIS OF A LOW COST CUSTOMIZED A-SCAN NON DESTRUCTIVE TESTING UNIT

M.N. M. Zukri and E. A. Bakar

[1]School of Aerospace Engineering,
Universiti Sains Malaysia,
*Email: meelmi@usm.my

ABSTRACT

This study introduced a new design of a low-cost customized A-scan Non Destructive Testing (NDT) unit using microcontroller-based scanning motion. The scanning unit from previous research has limitations in scanning envelope area and ability of detection. It can only detect large defects. To solve the limitations mentioned above, the purpose of this study is to develop a device with lower cost, more appropriate size of scanning envelop area, and ability to detect internal layer defections. The mechanism of this machine is achieved by using 3 NEMA 17 bipolar stepper motors for the movement of every axis. MATLAB is used in this machine for the purposes of signal acquisition, signal analysis and to control the linear motions. Since this machine is small in size (40cm x 27cm x 72cm), it can be used for small scale testing and analysis. As compared with current product, this machine is more portable and affordable. Thus, it is more convenient in the areas of research and education. The contributions of this machine is providing an affordable machine for advancement research field and enhancement quality inspection in manufacturing industries.

Keywords: Non Destructive Testing (NDT), scanning motion, signal acquisition, signal analysis.

INTRODUCTION

In advanced technology applications such as aerospace and industrial with emphasis on economics and safety, Non-destructive testing (NDT) is a practical method to validate and inspect the composite without compromising the integrity of the material. Commonly NDT techniques are Acoustic Emission Testing (AET), Acoustic Resonance Testing (ERT), Electromagnetic Testing (ET), Infrared Testing (IT), Magnetic Particle Testing (MT), Radiographic Testing (RT), Ultrasonic Testing (UT) and Visual Testing (VT). Ultrasonic Testing is the most widely technique used for the inspection of composite. [1] Currently, The existing product available for Ultrasonic Testing (UT) is quite expensive and the size is larger. Thus, it is inconvenient for the research and educational purpose. The previous researcher has develop low cost and small size of scanning unit. However, the scanning unit from previous researcher has limitations in scanning envelope area and ability of detection which can only detect the large defect. The aim of this work is to develop a device with lower cost, more appropriate size of scanning envelop area, and ability to detect internal layer defections. A high specification transducer which is 2.25 MHz immersion type is used to improve the capability of inspection to detect interlayer defect. The data acquire from hardware Arduino is analyses and the scanning result is evaluated. This data is evaluated by performing software using MATLAB programs.

Composite material is widely used in aircraft industries. In critical structural applications, it is more important for ensure the structural integrity during manufacturing composite structure. For composite materials, much of the NDT is conducted either during or immediately after manufacture of the component and the work is looking for defects. Usually, defect of composites can occur by impact, environmental effects or from unidirectional or cyclic load application, all of which have different effects on the defects already present or all different types of defect in one time. Thus, the complexity of the advanced composite materials manufacturing and

Composite in service maintenance represents challenges in developing optimized non-destructive tools and tests. Traditional metals based NDT methods are inappropriate and often misleading when applied to anisotropic and inhomogeneous composite materials. In advanced technology applications such as aerospace, it is critical to use and develop new practical composites NDT methods. Composite NDT in modified traditional and new tools consists ultrasonic, x- ray, acoustic emission, thermal, optical, electrical and a variety of hybrid methods. Ultrasonic pulse-echo is a well-established and widely used non-destructive testing technique. A pulse of ultrasonic energy, typically a few microseconds, is transmitted into the specimen in a direction normal to the surface. Figure 1 shows a typical pulse-echo set-up for a submerged immersion test.

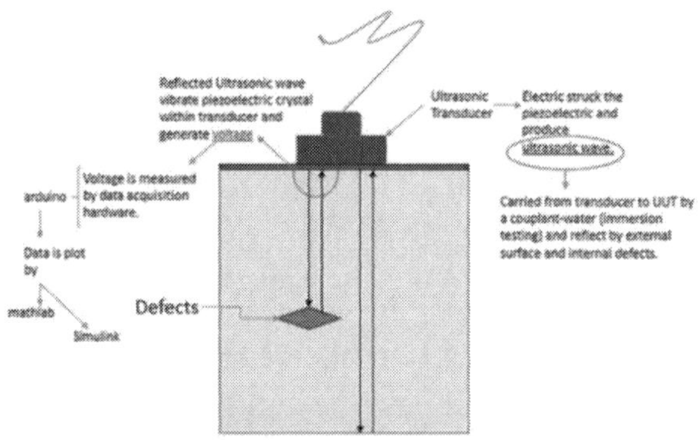

Figure 1. Immersion pulse-echo test with submerged specimen

Those signals which travel back towards the probe are detected and the position and size of a flow is determined from the total pulse travel time and detected amplitude respectively. This is the 'A-scan' display and it consists of a series of peaks, the position of which along the horizontal axis can be calibrated in terms of the depth in the composite. The amplitude of each echo will give some indication of the size and nature of the reflector, which might be a flaw or a specimen boundary. Figure 2 (a) shows a typical A-scan display from pulse-echo immersion test. In this display, the echoes from different features within the composite do not merge (i.e. they are well resolved) because the pulse duration is short to avoid it interacting with any of the features at the same time. The first peak is due to electrical pulse used to excite the transducer also called 'main bang' and is convenient reference for the following peaks. There are the flaw reflection in the front surface of the component and the rear surface of the component also known as back surface reflection.

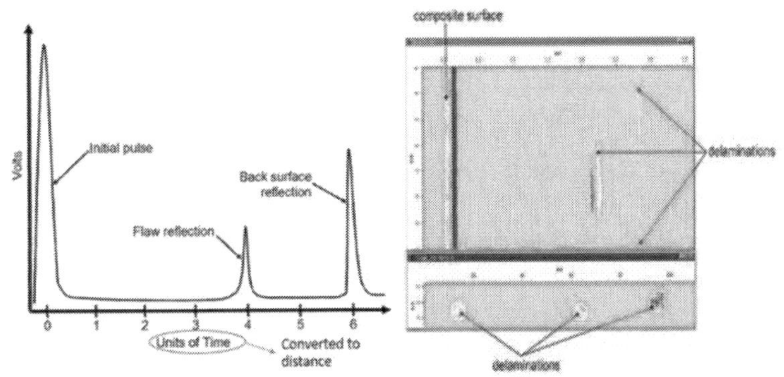

(a) (b)

Figure 2, (a) Diagrammatic A-scan display for a typical pulse-echo immersion test and (b)Ultrasonic image of delamination in carbon fibre composite (B and C scans)

Typically immersion testing is used to carry out area scans where the ultrasonic data can be represented in ways that make it easier to interpret. In the 'B-scan' display a 2D 'slice' through the specimen is produced by scanning the probe along the surface and capturing an A-scan at each point along the scanning direction. A 'D-scan' is also slice through the specimen taken at a fixed scanning point in each of a set of parallel B-scans. In B and D-scans, the amplitude at each point is displayed as a colour or a greyscale value according to a predetermined pallette. In general, the higher the amplitude the brighter or more intense the displayed point. C-scan may be thought of as 'Top down' or 'Plan' views of the component. Here the A-scan is 'gated' to display either the amplitude or the depth of signal occurring within a predefined depth range within the component. Figures shows a C-scan of a multi-ply CRP composite with artificially induced delamination (Teflon inserts). The upper image represents a B-scan and the lower image represents a C-scan as shown in figure 2(b).

METHODOLOGY

Design of Experiment

In this section, the experimental setup for project went further into the idea of improving Ultrasonic Testing machine for the general users, not only low-income ones. What started as a simple afterthought has now become a mission that must be completed without failure. A vision of an Ultrasonic Testing machine that is easy to use, light and cheap is produced. And thus, it moved on to improvise more on design. Morphological Chart is a table based on the function analysis. On the left side of the chart the functions are listed, while on the right side, different mechanisms which can be used to perform the functions listed are drawn, the flow of overall process is shown in figure 3(a), while the idea to compare alternative concept is shows in figure 3(b).

By through process design, the best alternative design 3 is selected as a final design. After that, fabrication process is start with dividing a parts of work. Some of works was completely done in mechanical workshop and the remaining is done in aerospace laboratory.

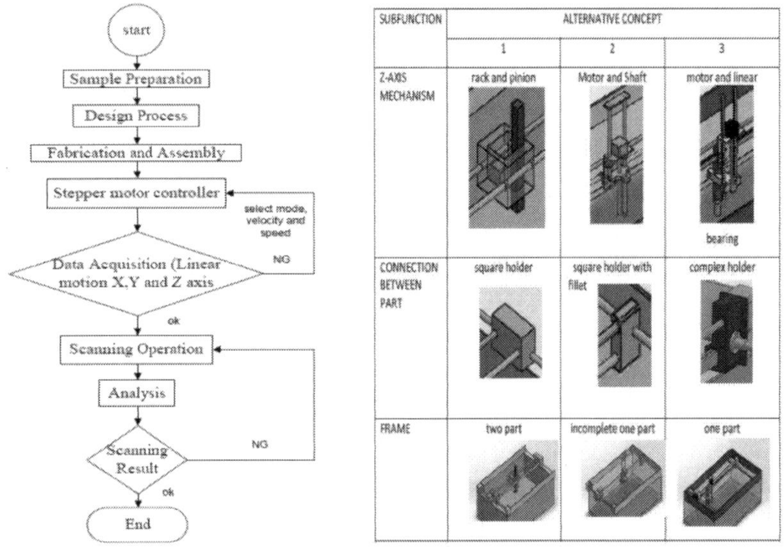

Figure 3. (a) Overall process flow (b) Alternative design

Motor Performance and Validation Test

The positional accuracy of the stepper motors will be high only when its step angle is very small. Hence for the inspection purpose, stepper motor is the best choice as it can offer small step angles in the range of 0.50 to 1.8 mechanical degrees. Thus, one of the experiment is set up to analyses the performance of motor. The experiment is to observe whereas the motor with or without loading influenced the speed performance of motor. One experimental is set up. The figure 4 (a) and (b) below shows the motor performance and validation test.

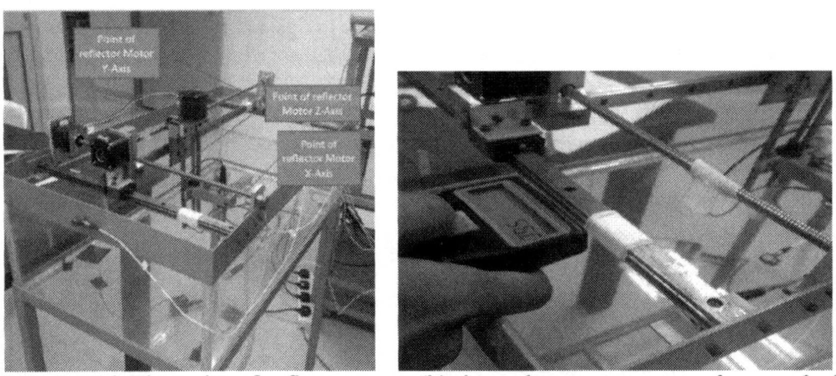

Figure 4. (a) The point of reflector for 3 axis (b) the tachometer measure the speed of motor X-axis

The reflector is attach to the thread. When motor is running, the reflector will rotate together and give the reading to the tachometer. The reading is taken as a data and efficiency of the stepper motor is calculated.

Fatigue Analysis thru analytical Approach for Motor

In this section, calculated the displacement, von-Mises (equivalent stress) (δ_{eq}), and factor of safety (f_s). Then, this section also to validate design stresses to compare meanstress value of material through graph as shown in figure 5. This curve is valid for steels. The aim of this analysis to prove the thread shaft instead of length, diameter and type of material is selected for infinite life. To design for infinite life, the endurance limit become the criteria of failure. At the end, the number of load cycles is calculate and the result should be exceed than $1x10^6$ cycles in order to withstand the infinite life of cycles. The example of travelling Y-Axis shown in table 1.

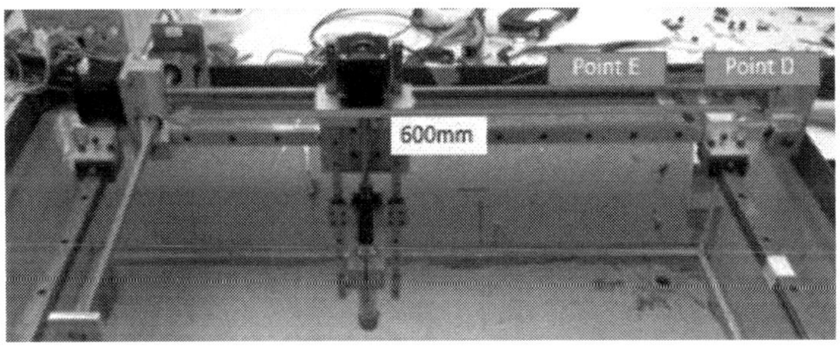

Figure 5. Example of Travelling motor 600mm

Table 1. Properties of Y-axis

Motors	Weight (N)	Length (m)	Diameter (m)	Yield Strength (N/m^2)
Y-Axis	9.04	0.60	0.008	172

Table 2. Analytical Approach

Variable (analytical approach)	Value
Displacement, ΔL	0.157mm
Von Mises /equivalent stress, δ_m	38.26 N/mm^2
Factor of Safety, f_s	1.72

The thread shaft is a rotating part that driven by motor is used to transmit power or motion. For the thread use in this machine, it consists 3 parts that associate with axis X, Y, and Z-axis. However, the critical part will be on the Y-axis thread because it is suffer the load from Z-axis and also the thread itself. Analytical Approach value is shown in table 2. For the thread on Y-axis, according to Shigley 8th edition [5], typically the static stresses are not significant in thread or shaft design because fatigue and speeds play a more important role in this design. Since, thread shafts is rotate, there is a strong likelihood of fatigue issues.

The fatigue life prediction is performed based on finite element analysis and analytical method. Using the constant amplitude loading, the fatigue life of the thread shaft has been predicted. This analysis will give information for the fatigue life of the thread shaft that are used in this machine. The comparison of simulation and analytical approach is presented in table 3.

Table 3. Comparison Simulation and analytical approach

Type	Simulation Approach	analytical approach	Percentage difference, %
Displacement, ΔL	0.101mm	0.157mm	36 %
Von Mises /equivalent stress, δ_m	41.51 N/mm^2	38.26 N/mm^2	8.5 %
Factor of Safety, f_s	>1 (blue color)	1.72	-

By comparing the above results, Von-Mises stress value by analytical approach δ_m=38.26 N/mm which are nearly close by using FEA approach having difference of 8.5% less than 10% which is acceptable range. The fatigue factor of safety calculated from analytical approach is 1.72 whereas the FEA approach give the FOS larger than 1. It is means that the design has a correct design and prevent to failure.

RESULTS AND DISCUSSION

Linear motion Analyses

Generally, to check the performance of motor, the experiment of speed rotation of motor is conduct as a mention in previous section. It is to checking the load that adding in particular point is affecting the performance of motor. The result is provides in table 4 and 5.

Table 4. Performance of motor without load

WITHOUT LOAD								
Motor	Volt	Current	Speed					Efficiency
			1	2	3	Average	Input	
X	5V	1.5A	56.3	57	56.3	56.53	200	28.27
Y			56.3	56.4	56.8	56.50	200	28.25
Z			56.4	56.3	56	56.23	200	28.12

Table 5 Performance of motor with load

WITH LOAD								
Motor	Volt	Current	Speed					Efficiency
			1	2	3	Average	Input	
X	5V	1.5A	56.7	56.5	56.6	56.60	200	28.30
Y			56.5	56.3	56.5	56.43	200	28.22
Z			56.4	56.5	56.1	56.33	200	28.17

It is clear from the above table 5 shown results that the performance of motors are not affected by the load attaching on the thread of the motor. The average speeds of all the conditions is nearly in range of 56-57 rpm. At last, the efficiency of the stepper motor roughly at 28 %. According to the previous researcher [4], the efficiency of the stepper motor is very low since the purpose for smooth motion and higher torque instead of speeding. So, it will show the load that attach on the thread of motor not effecting the performance itself.

Vibration structure Analysis

From the data collecting by the software of Integrated Measurement and Control (IMC), the Fourier Fast Transform (FFT) graph has created. The vibration of machine is analyze based on the FFT graph. In figure 6. To compare the effect of vibration on transducer by running motor Y-axis before the machine install the damper and after the machine install the damper. Constant Variables: Power Source 5V, mode of stepper motor (double mode) Manipulated Variables: Condition of machine (with / without damper)

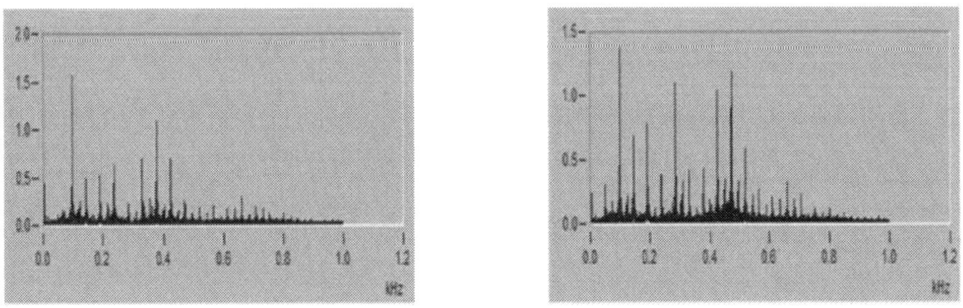

Figure 6. FFT graphs for motor Y-axis without damper (left) and with damper (right)

The aim of this work is to show the vibration on the transducer is reduce at the certain frequency by installing the damper. The transducer should prevent from vibrate in order to ensure the signal on the monitor is stable. The motor Y is selected as the source of vibration because this motor will constantly running during the scanning operation while the other motors is not critical.

By the observation on the FFT graph, it is showing the damping that install on the machine works at the certain frequency. At first, the desire frequency for motor X to reduce are 0.1 KHz and 0.38 KHz. The both frequency has the highest peak of amplitude which is 1.6ms^{-2} and 1.1ms^{-2}. After installing the damper, the both frequency amplitude is reduced to 1.4 ms^{-2} and 0.4 ms^{-2}.

Thus, the vibration at this frequency is successfully reduce. However, at the lower frequency, the damper is not work as it is practical that the damper is depend on frequency by dynamic stiffness.

Result of Inspection operation

In this project, the scanning operation is done by motor controller. The scanning is run to detect the defect at the various depth. Firstly, the depth is 2mm, 4mm and through hole. The transducer will travel on this path and provide the signal. Every depth give the different signal so it obviously shows the defect. In this operation, the speed is 200m/s,

200 step/rev. Every 1mm require 25 step. The length of this specimen is 128 mm, so the step require is 3200 step.

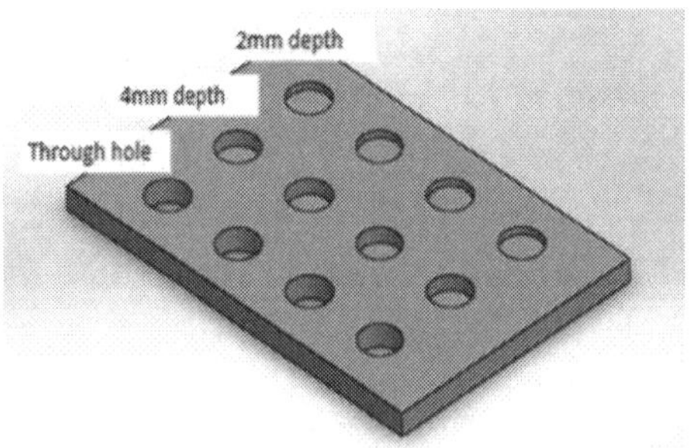

Figure 7 Isometric view of specimen

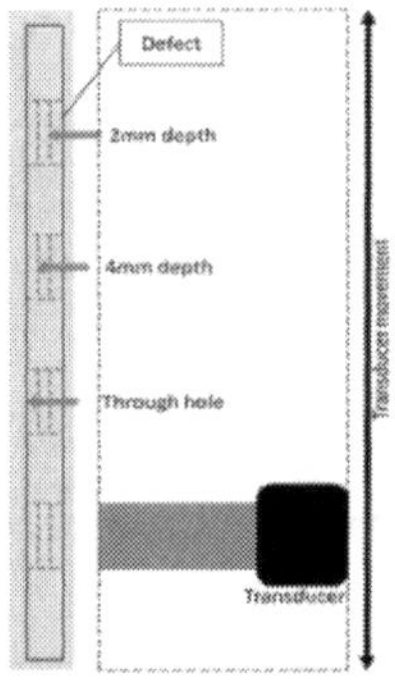

Figure 8 Cross sectional view of specimen

The figure 7 and 8 above shown the isometric view and the cross sectional view of the specimen. From the cross sectional view, it clearly show the different of depth that experienced by scanning of transducer. There is three different of depth where every depth has four holes. This scanning operation is done and the signal is shown in table 6. The simple sample is prepare for detect defect according to the various depth. In this scanning operation, for every defect that has trace by transducer, it will produce a sound as setting in program. So, it can clearly show that the defect was trace by the machine.

Table 6 Scanning operation

No	Transducer movement (Scanning operation)	Signal	Description
1	2mm depth		- The initial pulse and back surface reflection can obviously views - In the middle, the defect reflection can be observed close to initial pulse

CONCLUSION

In this study, the main objective is to develop a UT testing device with lower cost, more appropriate size of scanning envelop area, and ability to detect internal layer defections. The overall cost for develop this device is RM27 784.00 which is below to RM 30 000. For the current product, the cost of the device is usually above than RM500 000. Thus, lower cost is achieved. Then, the device also develop at size 40cm x 27cm x 72cm, so that the area of the scanning unit is become increase comparing with previous machine that has been developed in our lab. The mechanism of this machine is achieved by using 3

NEMA 17 bipolar stepper motors for the movement of every axis. MATLAB is used in this machine for the purposes of signal acquisition, signal analysis and to control the linear motions. For the purpose of ability in defect detection, the high specification transducer which is 2.25 MHz immersion type is selected for the capability of inspection to detect interlayer defect among composite laminates panel. Motion analyses together with vibration analyses is performs and the inspection result should the robustness of this machine to acquire data for Composite Panel as to be costumed made NDT machine.

ACKNOWLEDGEMENTS

The authors would like to acknowledge Universiti Sains Malaysia and Kementerian Pengajian Tinggi for providing the financial assistance under FRGS and PRGS grants (203/PAERO/6071296 and 203/PAERO/6740041) support for this project.

REFERENCES

[1] A. El Kouche and H. S. Hassanein, "Ultrasonic non-destructive testing (NDT) using wireless sensor networks," Procedia Comput. Sci., vol. 10, pp. 136–143, 2012.
[2] P. Cawley and R. D. Adams, "Defect types and non-destructive testing techniques for composites and bonded joints," Mater. Sci. Technol., vol. 5, no. 4, pp. 413–425, 1989.
[3] I. G. Scott and C. M. Scala, "A review of non-destructive testing of composite materials," NDT Int., vol. 15, no. April, pp. 75–86, 1982.
[4] M. F. Mahmod, M. Z. M. Pauzi, and E. A. Bakar, "Flatbed Scanner Image and Single Ultrasonic Defect Detection," no. November, pp. 26–27, 2013.
[5] S. R. Doctor, T. E. Hall, and L. D. Reid, "SAFT —the evolution of a signal processing technology for ultrasonic testing," NDT Int., vol. 19, no. 3, pp. 163–167, 1986.
[6] P. Gaudenzi, M. Bernabei, E. Dati, G. De Angelis, M. Marrone, and L. Lampani, "On the evaluation of impact damage on composite materials by comparing different NDI techniques," Compos. Struct., vol. 118, pp. 257–266, 2014.
[7] B. Djordjevic, "Nondestructive test technology for the composites," … Slov. Soc. non-destructive Test., pp. 259–265, 2009.
[8] B. Ren and C. J. Lissenden, "Ultrasonic guided wave inspection of adhesive bonds between composite laminates," Int. J. Adhes. Adhes., vol. 45, pp. 59–68, 20

CALCULATING THE MODULI ELASTICITY FOR REINFORCED CONCRETE USING NEW RULE OF MIXTURES APPROACH FOR THE DAM STRUCTURE

Zainordin Firdaus Zulkefli[2], Mohd Hafiz Zawawi[1], Nurul Husna Hassan[2], Mohd Remy Rozainy Mohd Arif Zainol[2*], Mohamad Aizat Abas[3] and Ahmad Zhafran Ahmad Mazlan[3]

[1]Department of Civil Engineering, College of Engineering, Universiti Tenaga Nasional, Selangor,
[2]School of Civil Engineering, Engineering Campus, Universiti Sains, Malaysia, Penang
[3]School of Mechanical Engineering, Engineering Campus, Universiti Sains, Malaysia, Penang
[*]Email: ceremy@usm.my

ABSTRACT

Dam is the important structure use for water domestic store, electricity power supply, irrigation and flood control. Dam deal with reinforced concrete as a main material in construction. The material considerable safe when ability to support external and internal load. The capability of dam material due to safety is effect by degradation. Prevention monitoring must be conducted for high risk structure. Thus, moduli elastic is a mechanical property to measure stiffness and had relationship between stress and strain of material. While non-destructive and destructive testing in site inspection does not include reinforce and give incorrect moduli elasticity. New approach rule of mixture (Rom) with solid work software refer to United States Department of the Interior Bureau of Reclamation (USBR) standard design for small dam, give reliability monitoring with true sustain modulus elasticity of dam structure time by time. The concrete properties for sustain of moduli elasticity standard type gravity concrete dam value is 20.684 GPa. If the value decreases drastically from the standard, high precaution must be taken.

Keywords: Dam, safety, prevention monitoring, reinforced concrete, destructive, non-destructive, rules of mixture (Rom), solid work, and USBR standard and moduli elasticity.

INTRODUCTION

Dam are infrastructure provide service to public for long term with economy value. Expend service of life these infrastructures with preventive and corrective maintenance continuously. The dam are consider in failure mode if cannot containment the water at upstream. Dam failure can cause fatality, property damage, water supply disruption and other negative impact. A dam must construct with concrete material where ability to sustained loading, thus concrete properties must have criteria such as strength, permeability, durability and other properties [1]. Therefore, the dam must always put in safety condition. Monitoring can be done toward observation and inspection. Destructive tests are the method where the sample or specimen is crushing to get information about strength and stress strain by using mechanical machine. Destructive testing using small concrete core with diameter 50 mm because it gives very small damage and avoid cutting the reinforcement of the structure [2]. Meanwhile non-destructive test where the devices is use directly to the structural to estimate the strength of concrete without damage the structure [3]. This study was conducted to determine reliable sustain moduli elasticity using new approach rule of mixture, difference between moduli elasticity (new approach) and moduli elasticity (normal practice). Minimum value for dam concrete material properties refer to United States Department of the Interior Berau of Reclaimation Design of Small Dams (USBR) shown in Table 1.

Table 1. Clause 8.4 (b) criteria concrete material for dam [4]

Compressive strength (static): 3,000 to 5,000lb/in^2 (20.68427 N/mm^2 -34.47379 N/mm^2)
Tensile strength (static): 5 to 6 percent of the compressive strength
Tensile strength (dynamic): 10 percent of the static compressive strength
Shear strength (static):
Cohesion: 10 percent of the static compressive strength
Coefficient of internal friction: 1.0
Poisson's ratio: 0.2
Sustained modulus of elasticity: 3.0×10^6 lb/in^2 (20.684Gpa)
Coefficient of thermal expansion: 5.0×10^{-6}ft/ft/°F
Unit weight: 150 lb/F^3
Diffusivity: 0.05 ft^2/hr

As shown in Table 1, compressive strength is between 20.684 N/mm^2 - 34.474 N/mm^2. While sustained modulus elasticity is 20.684 GPa. Rule of mixtures is suitable formulae for composite material such reinforcement concrete in term to determine modulus elasticity [5]. The standard practice not counting reinforcement value to determine sustain moduli elasticity. Modification approach of Rules of mixture (Rom) is to get true value of moduli elasticity for current dam monitoring.

METHODOLOGY

Experimental Procedure

If the dam necessary to core use destructive test. Core the dam structure (50 mm diameter) for the section had a problem or critical and label the location of specimen, then cutting the specimen L/D = 2 (Standard) [6], measure the diameter and height of specimen, next key in the value diameter and height of specimen into compression software, setup the graph stress strain, lastly start to compress until sample crush and failure. Provide stress strain graph results for next step for modulus elasticity calculation. If the risk is too high dam core is not aloud, use nondestructive to produce stress strain results for modulus elasticity calculation.

Moduli Elasticity Calculation

The E value without reinforcement can be calculated using the following equation:

$$E_{\text{normal practice}} = \sigma / \mathcal{E} \tag{1}$$

$E_{\text{normal practice}}$ is without reinforcement, concrete only σ / $\mathcal{E}$ (From compression test results)

Rule of mixtures (Rom) [7]

$$E_{\text{new approach}} = [(E_{\text{concrete}})\,(V_{\text{concrete}}/\,V_{\text{total}})] + [(E_{\text{steel}})\,(V_{\text{steel}}\,/\,V_{\text{total}})] \tag{2}$$

E_{concrete} = modulus for concrete
E_{steel} = modulus for reinforcement
$E_{\text{new approach}}$ = combination concrete with reinforcement
V_{concrete} =Volume of friction of concrete from solid work software
V_{steel} = Volume of friction of reinforcement from solid work software

Volume friction of reinforcement can be determined by draw back reinforcement into SolidWorks software refer to original as built drawing. Select the critical path of the dam. SolidWorks can generate volume information shown in Figure 1.

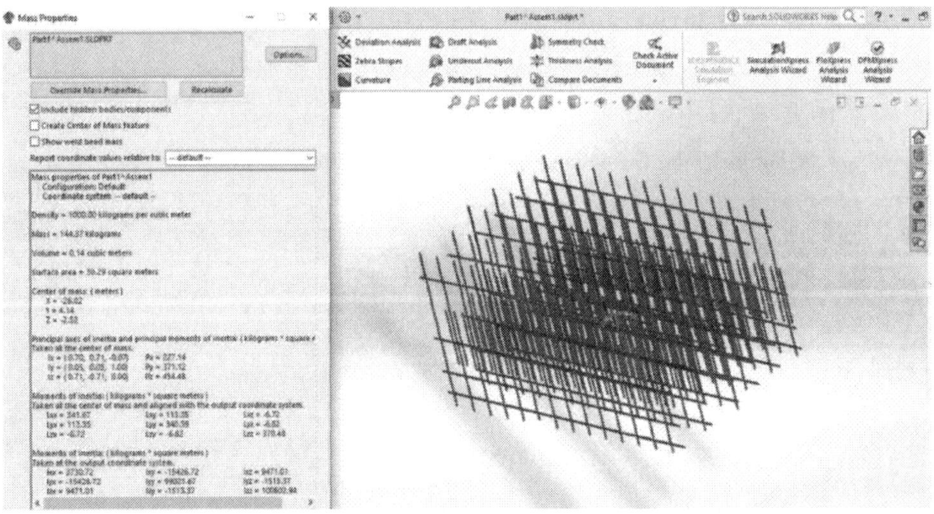

Figure 1: SolidWorks generate volume friction of reinforcement from the drawing.

SolidWorks give detail output volume of friction for reinforcement and details of information in mass properties shown in Figure 2

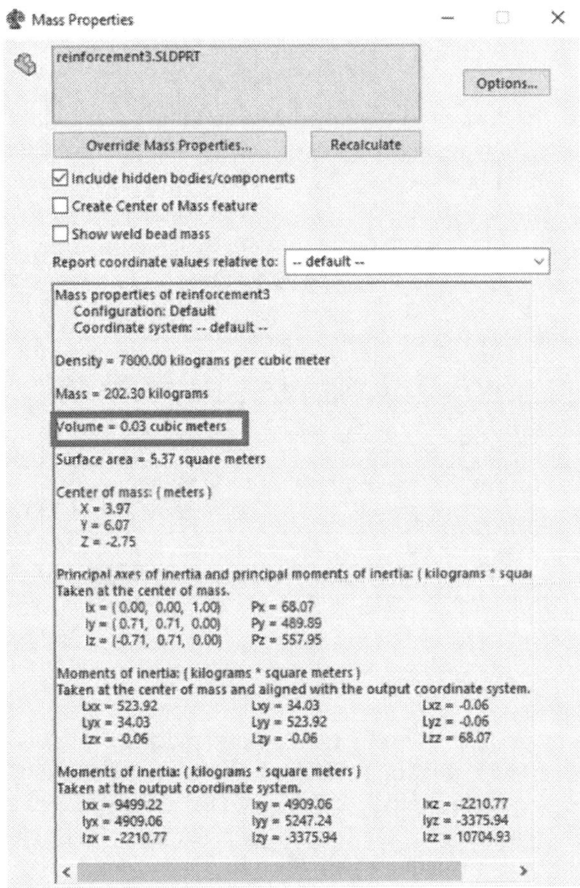

Figure 2: Mass properties of SolidWorks.

Volume friction of concrete also can be defined, draw back shape part of dam in solid work software shown in figure 3, insert density value of concrete with correct ratio. Make sure that all dimension is correct.

Figure 3. Volume friction of concrete in SolidWorks software

RESULTS AND DISCUSSION

In this section, the result show difference between moduli elasticity (normal practice) and moduli elasticity (new approach) shown in Table 2. The higher difference is 5.41 GPa and the lowest difference is 3.65 GPa. The average of difference is 4.45 GPa.

Table 2. Difference between moduli elasticity (normal practice) and moduli elasticity (new approach).

Specimen	Moduli elasticity (Normal practice), GPa	Moduli elasticity (New approach), GPa	Difference, GPa
1	21.94	26.81	5.41
2	21.17	25.73	4.56
3	20.11	24.89	4.78
4	20.13	23.78	3.65
5	19.14	22.97	3.83

CONCLUSION

In this study, the main objective to determine true value moduli elasticity using new approach of Rule of mixture (Rom). Table 3 shows result of moduli elasticity new approach compare to sustain moduli elasticity USBR(standard). All specimen are not exceed 20.684 GPa. The difference must not ignore because the value in GPa consider big. If the value estimating is over, it will affect the cost of material such as decrease of bar reinforcement, decrease size of bar, decrease of mixing ratio of concrete. If the estimating value is less, the situation always in failure but reality condition is safe. The best practice is getting the true value. Monitoring can be done accurately.

Table 3. Result of moduli elasticity new approach compare to sustain moduli elasticity USBR (Standard).

Specimen	Moduli elasticity (New approach practice), GPa	USBR Sustain moduli elasticity, GPa
1	26.81	20.68
2	25.73	
3	24.89	
4	23.78	
5	22.97	

REFERENCES

[1] Bakenaz A. Zeidan(2014). "Design and analysis of concrete gravity dam".Article Review.Tanta University.

[2] Yamamoto, H. Hamada, Y. Sagawa, and T. Hiromitsu, "Evaluation of compressive strength of concrete using small diameter core."

[3] Samia Hannachi and M Nacer Guetteche. Review of the rebound hammer method, estimating concrete compressive strength on the site.

[4] United States Department of the Interior Berau of Reclaimation Design of Small Dams (USBR).

[5] G. R. Liu, "A step-by-step method of rule-of-mixture of fiber- and particle-reinforced composite materials," Composite structure vol. 40, pp. 313–322, 1998.

[6] S. Suzuki, T. Kage, and S. Seko, "Influence that Ratio of Length to Diameter of High-Strength Concrete Core to Compressive Strength of Concrete," pp. 1–8.

[7] A.S. Ezeldin, P.N. Balaguru, Normal and high-strength fiber-reinforced concrete under compression, Journal of Materials in Civil Engineering 4 (4) (1992) 415–429.

TOPOLOGY ANALYSIS OF BICYCLE RIM BRAKE PAD TO IMPROVE BRAKING PERFORMANCE

Le-Onn Keong[1] and Choe-Yung Teoh[1*]

[1]Faculty of Engineering,
Tunku Abdul Rahman University College.
*Email: teohcy@tarc.edu.my
Phone: +60341450123; Fax: +60341423166

ABSTRACT

Various parameters of rubber brake pad tribology will affect the braking performance of a rim brake system of a bicycle. Out of those, three main parameters are contact length of brake pad, surface topology of friction surface, and the thickness of rim brake pad. The goals of this study are to improve the braking performance of rim brakes to have better friction performance while retaining its simplicity of manufacturing and to investigate the effect on friction performance through the modifications of external dimensions and surface topology of rim brakes. Moreover, through this study, it can reveal which topology modifications of the rubber brake pad will have the most significance effect towards friction performance. Experimental measurement is used to obtain the friction properties and then parametric study was carried out numerically to obtain the braking performance of the rubber brake pad. The parametric study was simulated in ANSYS Transient Structural analysis. Taguchi design method was used for quantitatively identifying the right inputs and parameter levels. The results showed that smaller groove and smaller width contribute to higher friction performance. Stress concentration occurs at the edge of groove; hence, by reducing number of groove, it allows higher braking force to be generated. Location of groove concentration appears to have insignificance effect to the friction performance. Through the topology modifications, it improves the braking performance by 3%. As for external dimension, thick rubber brake pad will not demonstrate any improvement and longer contact length will demonstrate higher friction force. However, it has to be bounded by the practicality of the dimension. Upon modifications at the external dimension, it has an improvement of 64% in braking performance.

Keywords: Rubber brake pad, friction force, coefficient of friction, surface topology, tribology, groove pattern

INTRODUCTION

Braking performance of a bicycle is important as it is a crucial mechanism to stop the bicycle in an emergency event. In order to obtain a higher performance brake pad, it requires brake pad materials that are made from various combinations of high-end braking material and it will increase the price of the brake pad. Hence, most users will prefer not to purchase high performance brake pad due to the cost. Braking performance of a conventional rim brakes (rubber brake pads) is not as efficient as a standard disc brake system. Commonly available rubber brake pad (RBP) in the market has low friction and low braking performance. The poor performance leads to lack of power in braking and as a results, slippage and sliding may occur during braking.

The main objectives of this research are to improve the performance of rim brakes to have higher friction performance through parametric study, to investigate the effect on friction performance through the modifications of external dimension of rim brakes, and to analyse the effect of different extent of tribology features of rim brake pad acted on the rim. The contribution of the research is to improve the braking performance and friction performance of a rim brakes in to improve the safety of braking system without the necessary to increase additional cost by putting the interest of the study on the tribology properties and dimension alteration. The study is going to carry out fully on rim brakes or rubber brake pads, whereas many previous studies are completed only to further improve disc brake.

METHODOLOGY

I. Experimental Test Bench Setup

In order to obtain the coefficient of friction between the rubber brake pad and aluminium rim, a test bench is constructed (Figure 1). The load cell used is an S-type structure (DYLY-103) that is suitable for tension and compression. It has the sensitivity of 2.0 ± 0.005 mV/V. The load cell is connected to a digital reader.

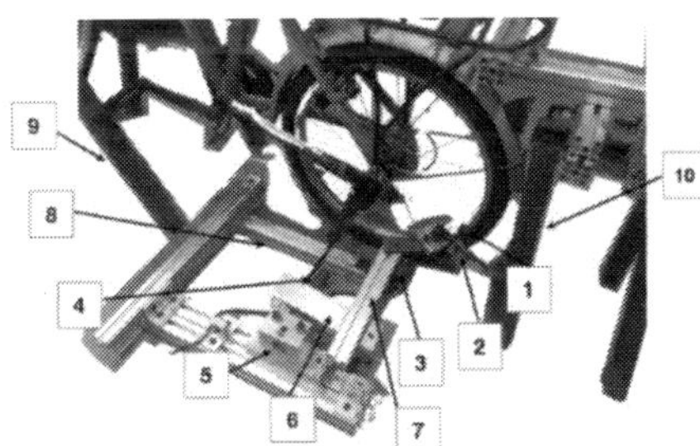

Figure 1. 1.Rubber brake pad, 2.Brake lever, 3.Load cell setup, 4.Shaft, 5.Bearing housing with bearing, 6.Aluminium profile, 7.Longitudinal aluminium profile, 8.Support aluminium profile, 9.Front stand, 10.Back stand

II. Numerical Simulation Setup

Reaction forces on the surface of contact, deformation at groove edge, and shear stress generated at the surface of contact are captured and recorded from the ANSYS simulation results. Figure 2 shows the 3D model used in the study.

III. Parametric Study of Rubber Brake Pad Model

Rubber brake pad will undergo parameters change to further improved braking performance by altering a few crucial parameters that could affect the braking efficiency of the rubber brake pad.

Changes of Contact Length towards Friction Force Performance

The length of the brake pad is varied and the effect on the braking performance respond to the length variation is studied. This research will analyse the suggested length of brake pad as stated in Table 1.

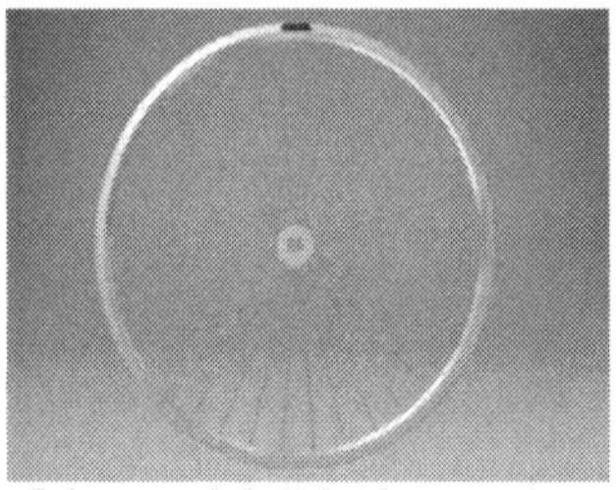

Figure 2. Front view of the model. Black represent rubber brake pad and aluminium represent rim.

Table 1. Contact lengths that will be used to determine the best contact length

Set	Length (mm)
1	54
2	57.4
3	70

Alteration of Groove Pattern Topology towards Friction Force Performance

A number of properties of groove pattern topology are modified in this study. Groove depths are varies between 3 mm to 9 mm, number of straight grooves are varies between 3 to 6, groove widths are varies between 3 mm to 7 mm, and location of grooves are separated to symmetrical, leading edge, and trailing edge.

Changes of Thickness of Brake Pad towards Friction Force Performance

A thicker brake pad allows more wear and deeper groove tolerance since it has the allowable thickness to be machined but tends to cause more vibration to occur when load applied. Suggested thickness of brake pad are 14 mm, 20 mm, and 25 mm for baseline, set 1, and set 2 respectively.

Materials Properties of Rubber Brake Pad and Aluminium Rim

The material properties of the aluminium rim and rubber brake pad for both experimental measurement and numerical simulation (Table 2) has to be matched as close as possible to reduce inaccuracy in reading.

Table 2. Materials parameters used during the numerical analysis (Shackelford et al., 2016) (Ramezani et al., 2012)

	Aluminium Rim	Rubber Brake Pad
Young Modulus (GPa)	70	0.015
Poisson's Ratio	0.33	0.499
Density (kg/m^3)	2700	1500

RESULTS AND DISCUSSION

The tribology properties between a rubber brake pad and an aluminium rim are represented by the coefficient of friction (COF). The COF is determined at the

https://doi.org/10.1051/matecconf/201821704004

experimental stage where a custom bench is fabricated to harvest the applied normal force (Table 3) and resultant friction force during the braking event.

I. Experimental Measurement of Normal Force and Friction Force

The properties between both of the surfaces is well portrayed by the COF. Thus, the numerical simulation will then use the obtained COF (Table 4) during the setup in order to mimic the surface condition of the rubber brake pad and aluminium rim without necessary of detailed study on the surface.

Table 3. Applied force obtained during execution of braking process

Trial	Applied Normal Force (N)
1	118
2	106
3	107
4	109
5	92
Average	106.4

II. Numerical Analysis on the Friction Performance

First set of simulation (Stage 1) results is prepared and run by varying four major topology parameters of a rubber brake pad using Taguchi's Method. Table 5 shows the friction force generated at the effective contact surface between the bicycle aluminium rim and the surface of the rubber brake pad. The deformation is labelled as "DEFORM" and its results are highly inaccurate. Rubber brake pad configuration of "333A" has the largest friction force among the configurations. It has groove depth of 3 mm, 3 straight grooves, and 3 mm groove width. The ranking then followed by 533A and 333C.

Table 4. Friction force obtained from the reading of the load cell

Trial	Friction Force (N)	Coefficient of Friction
1	39.0	0.3665
2	44.0	0.4135
3	42.0	0.3947
4	42.0	0.3947
5	46.0	0.4323
Average	42.6	0.4004

Table 5. Friction force obtained from numerical analysis during Stage 1

Set	Groove Depth (mm)	Number of Straight Grooves	Groove Width (mm)	Friction Force (N)		
				Location of Grooves Concentration		
				Symmetrical (S)	*Trailing Edge (A)*	*Leading Edge (C)*
1	3	3	3	28.27	29.09	28.33
2	5	3	3	26.75	28.37	28.12
3	7	3	3	24.82	DEFORM	DEFORM
4	9	3	3	23.26	DEFORM	DEFORM
5	3	5	3	24.95	24.98	24.78
6	5	5	3	24.94	23.19	24.25
7	7	5	3	DEFORM	DEFORM	DEFORM
8	9	5	3	DEFORM	DEFORM	DEFORM

9	3	3	5	DEFORM	24.01	17.45
10	5	3	5	21.96	23.96	24.72
11	7	3	5	20.79	25.15	DEFORM
12	9	3	5	19.92	DEFORM	DEFORM
13	3	3	7	19.35	19.67	20.44
14	5	3	7	17.98	20.64	20.24
15	7	3	7	17.85	DEFORM	DEFORM
16	9	3	7	DEFORM	DEFORM	DEFORM

III. Friction Performance Respond on Groove Width Variation

Compared to a 5 mm groove width, the side walls of a 3 mm groove width are well supported and the fillet from the base of the groove act as a buttress for the side walls (Consolacion et al., 1995).

IV. Friction Performance Respond on Number of Straight Groove (Groove Density)

Stress concentration exist when there is discontinuity in geometry. Higher number of grooves demonstrates a lower shear stress exist at the edge of each groove (contact point during deformation of groove) (Nakashima et al., 2009).

V. Significance Location of Groove Concentration

The results demonstrates the position where the concentration of grooves exist do not affect the results significantly. The variations between the different location for the same series are small (~4% variation) and it is concluded that it is not the dominance feature in a rubber brake pad. One-way ANOVA test results also shows the feature do not give a significance score ($p < 0.05$) towards the friction performance of the rubber brake pad.

Depth of Groove on Friction Performance

From multiple results obtained, it is cleared that most groove depth that is more than 5 mm of depth will suffer deformation and failure upon loading by the surface friction. While for 3 mm and 5 mm, groove depth of 5 mm tend to have slightly higher flex when it is loaded, as seen in Figure 3.

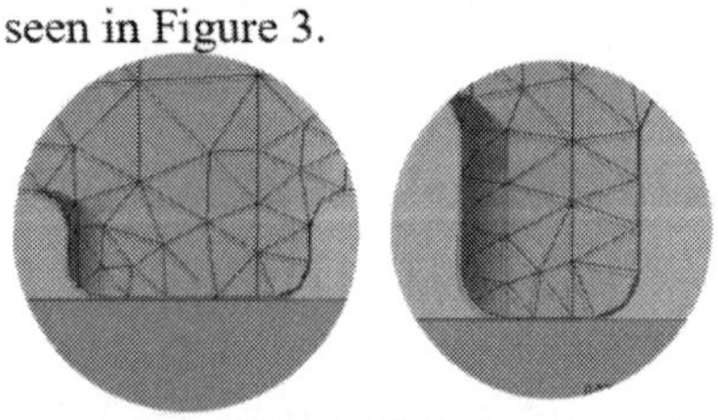

Figure 3. One of the contact point of "353A" (left, enlarged-view) and "553A" (right). The "553A" shows a higher flex to the left hence reducing the effective contact area on the right.

In second stage of numerical analysis, it shows larger contact length will contribute to higher friction force performance. This is largely due to larger area of surface rubbing on the surface of the rim. However, in practicality of manufacturing and usage, it is not advisable as it will always limits by the radius of the rim and the cost to manufacture as it requires more raw materials for a large rubber brake pad. Apart from that, thicker rubber brake pad will not contribute to any better friction force performance, even for the same contact length. The findings is in-line with what Smith

(2011) and Sanches et al. (2017) demonstrates in their research where it shows no noticeable increment in friction force when thickness of rubber increases.

Hence, all in all, it reveals that longer contact length with the same thickness as baseline model or RBP: 333A-CA, demonstrates the highest friction force performance.

CONCLUSION

For a rubber brake pad topology features, not all features give a significance changes to the friction performance. Through the numerical analysis, it demonstrates different changes for different topology features;

a) Smaller groove depth will result in higher friction force due to lesser flex occurred at the groove edge; hence it will not reduce the contact area when it is subjected to loading.
b) Three number of grooves perform better than five number of grooves. This is due to stress concentration exist when there is groove due to geometric discontinuity.
c) Smaller groove width demonstrates higher friction force too due to the geometric support at the base of the groove.
d) Location of groove concentration does not give significance difference in the friction performance.

For dimensional changes, larger contact length will result in higher friction performance and thicker rubber brake pad will result in no negative friction performance. Comparing with the baseline model (333S), the modified version of rubber brake pad during Stage One improved by 3% (333A), and Stage Two modification improved significantly by 64% (333A-CA) in terms of friction force performance

ACKNOWLEDGEMENTS

I would like to thank Tunku Abdul Rahman University College (TARUC) for organising such hands-on oriented final year project. Besides that, I would love to say a million of thank you to my supervisor, Dr. Teoh Choe Yung, for his unconditional supports throughout this project. Lastly, I would like to express my fullest gratitude and appreciation to all of my family members, especially for my parents for their wholehearted supports and concerns.

REFERENCES

[1] Consolacion, R.E., Brayer, R.R. and Croyle, W.L., The Goodyear Tire & Rubber Company, 1995. *Tread for a tire*. U.S. Patent 5,407,005.

[2] Nakashima, H., Takatsu, Y., Shinone, H., Matsukawa, H. and Kasetani, T., 2009. FE-DEM analysis of the effect of tread pattern on the tractive performance of tires operating on sand. *Journal of Mechanical Systems for Transportation and Logistics*, *2*(1), pp.55-65.

[3] Ramezani, M. and Ripin, Z.M., 2012. *Rubber-pad forming processes: Technology and applications*. Elsevier.

[4] Sanches, A.O., Kanda, D.H.F., Malmonge, L.F., da Silva, M.J., Sakamoto, W.K. and Malmonge, J.A., 2017. Synergistic effects on polyurethane/lead zirconate titanate/carbon black three-phase composites. *Polymer Testing*, *60*, pp.253-259.

[5] Smith, R.H., 2011. A Load-Independent, Adhesion-Related Rubber Friction Force?. *The Journal of Adhesion*, *87*(11), pp.1112-1123.

EFFECT OF STEP HEIGHT ON THE AERATION EFFICIENCY OF CASCADE AERATOR SYSTEM USING PARTICLE IMAGE VELOCIMETRY

A. Azman[1], M. H. Zawawi[2], N. H. Hassan[2], A. Abas[1*], N. A. Razak [3], A. Z. A. Mazlan[1] and Mohd Remy Rozainy M.A.Z [4]

[1]School of Mechanical Engineering, Engineering Campus, Universiti Sains Malaysia, Penang
[2]Department of Civil Engineering, College of Engineering, Universiti Tenaga Nasional, Selangor
[3]School of Aerospace Engineering, Engineering Campus, Universiti Sains Malaysia, Penang
[4]School of Civil Engineering, Engineering Campus, Universiti Sains Malaysia, Penang
[*]Email: aizatabas@usm.my

ABSTRACT

Aeration is an important parameter in water filtration system as it allows the transfer of oxygen to water through turbulence effect which subsequently increases air entrainment in the water. For water treatment application, aeration efficiency is measured to ensure continuous re-oxygenation of the unfiltered water. Aside from aeration, this paper also studies the flow patterns through the use of particle image velocimetry (PIV) setup. Through the use of real scale down physical model laboratory study is performed using PIV to obtain the velocity profile. These velocity profiles will then be used to calculate the aeration efficiency of the water in a cascade aerator system. Based on the findings, the aeration efficiency obtained from the PIV experiment has a maximum value at the lowest point of the cascade aerator system with a value of 0.0139 due to increase in mass flow rate as it moves through the steps with velocity of 0.418 m/s. Therefore, in the design of a cascade aerator system, it is advisable to increase the number of steps since it will increase the aeration efficiency of the system.

Keywords: PIV, Aeration, Cascade aerator, water treatment, water filtration.

INTRODUCTION

Cascade aerator is often linked with the stepped spillway system due to its similar configurations and purposes. As an application of the stepped spillway which high potential air-water gas exchangers [1], cascade aerator is used for in-stream re-aeration and it is widely used in water treatment plants to enhance the air-water transfer of atmospheric gasses and the removal of iron and manganese which can cause unpleasant taste and odors. In a nutshell, the basic idea of a cascade aerator is to generate aeration which is extremely helpful in facilitating the transfer of gas into or out of the water. The typical design of a cascade aerator is shown as in Figure 1 below.

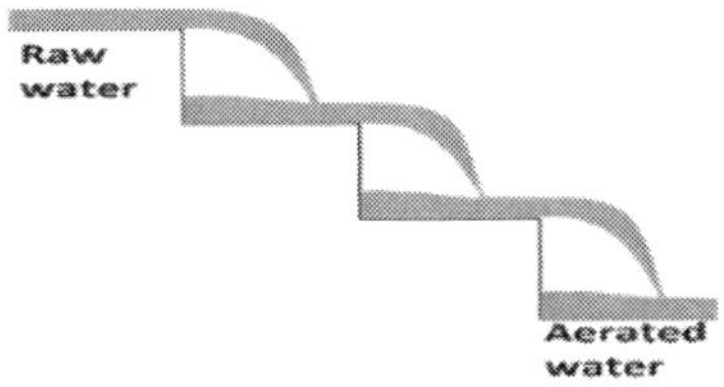

Figure 1: Cross-sectional side view of cascade aerator [1]

Various studies have been concerted to improve the performance of cascade aerator system. In a study by Amador et al, PIV method is used to compute the fluctuation in velocity that is expressed in terms of root-mean square (rms) of the instantaneous velocities. The group found that the maximum values are found to be near to the pseudo-bottom [2]. In another study, velocity fields were measured by Large Scale Particle Image Velocimetry (LSPIV) and the results showed that relative flow velocity is slightly higher for greater spillway size [3]. Air water flow properties comprise of the comparison in interfacial velocity, bubble count rate and turbulence intensity between the two spillways with different slopes. The comparative analysis proposed that smaller residual energy (better energy dissipation) happens at pooled stepped spillway with slope of 8.9^0 [4]. Another study also prove that the capabilities of PIV indicates velocity fields and profiles were highly organized although velocity of rotating impeller was high [5]. In addition, the velocity distributions inside and outside of droplet were measured and result showed that a pair of symmetrical vortices existed during droplet formation process in their experimental work [6]. In this work the aerator efficiency was obtained based on the step height of cascade aerator.

METHODOLOGY

Modelling of Cascade Aerator

A river bank model was drawn with the aid of powerful and easy-to-use Solidworks 3D CAD software. The individual parts of the river bank with exact dimensions were formed by considering the complexity of the geometry. The river bank model is then created via Solidworks and fabricated as shown in Figure 2 below.

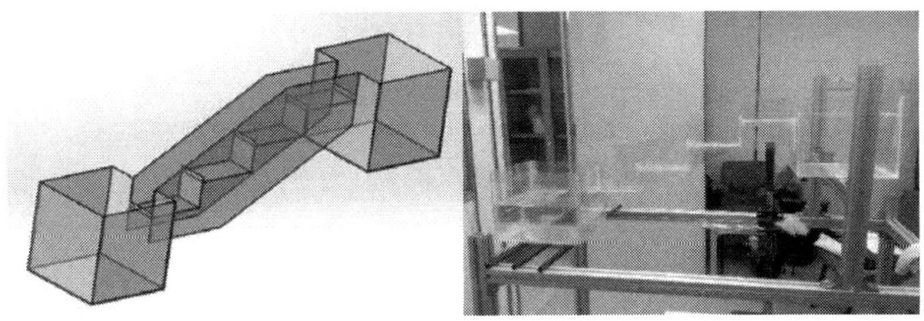

(a) (b)

Figure 2: (a) CAD model Configuration of set 1 model with barrier height 10mm, (b) Fabrication of the Model

Experiment Setup

The measurement of water velocity could be done using PIV illumination system which is an optical method to determine the instantaneous vector measurements in cross section flow. Velocity vector of the fluid is calculated and determined from the movement of the tracer particle between two light pulses:

$$V = \frac{\Delta x}{\Delta t}$$

where Δx is the displacement of a marker, located at x at time t, over a short time interval Δt separating observations of the marker images. Therefor the changes of velocities can be compute with frame by frame. The video will extract as picture frame by frame and then transfer to PIVlab

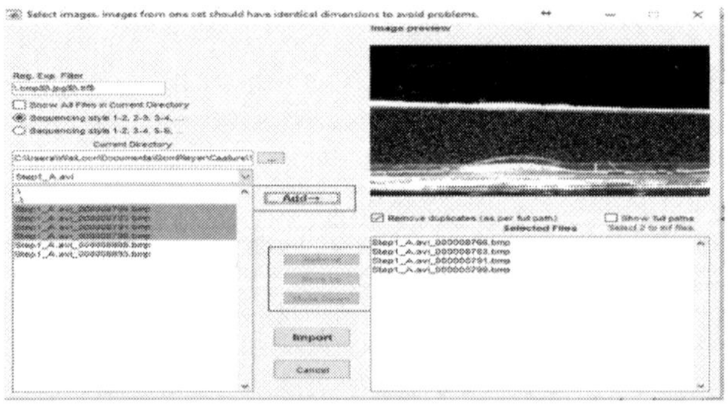

Figure 4: Importing of images to be used in the screen panel of PIVlab

Several crucial materials and instruments would have to be prepared and then set up properly before the carrying out the experiment. The above-mentioned apparatus is then installed properly and the actual arrangement of apparatus for two different configurations of river bank model are shown in Figure 5.

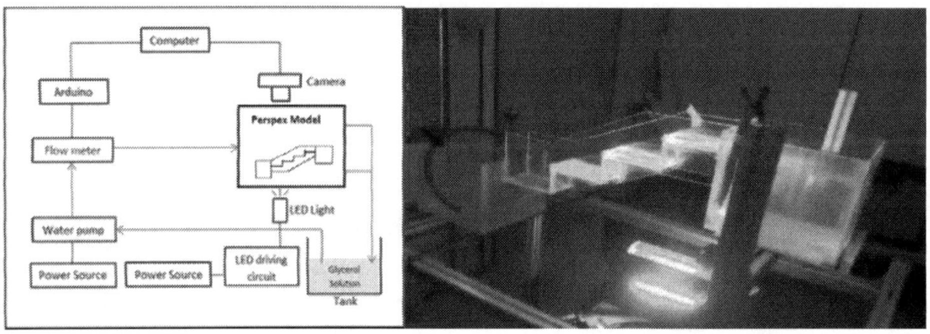

(a) (b)

Figure 5: (a) Schematic Diagram of experiment setup (b) Actual experimental setup for model with interior lighting being turned off

RESULTS AND DISCUSSION

The visual investigations of the flow patterns included the observations of the water flow patterns for all river bank configurations for one specific discharge rate. The configuration exhibits a nappe flow regime as depicted in Figure 6 with typical flow patterns of free-falling jets from one step to the next step. Snapshots of the simulated flow patterns are shown in Figure 7 which illustrates the velocity vector profile.

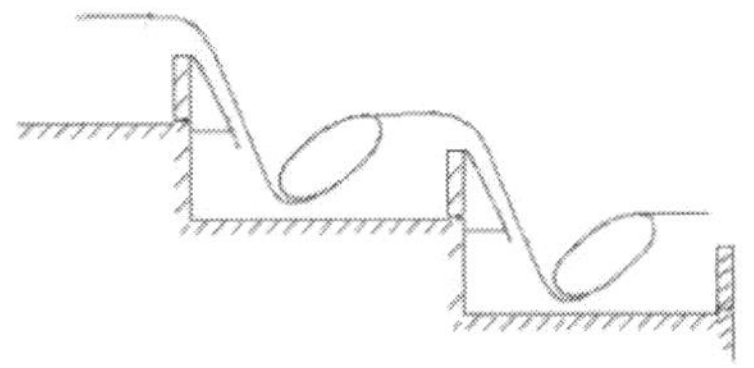

Figure 6: Flow over pooled stepped cascade

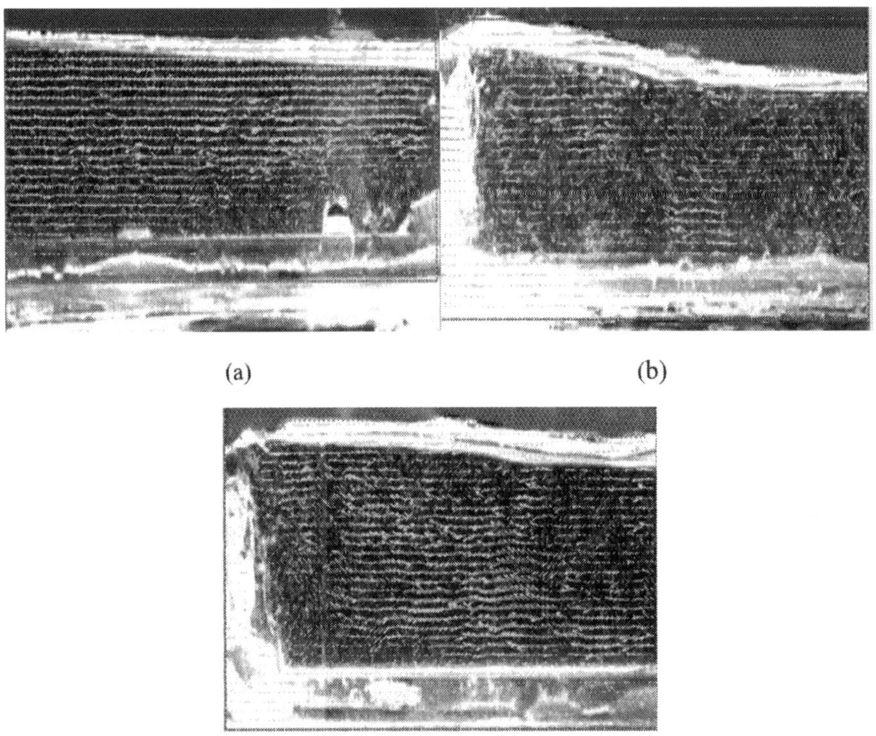

(a) (b)

(c)

Figure 7: PIV flow profiles across different steps for (a) First step (b) Second step, (c) Third step

PIV experiment compute velocity variations across the different stepped spillway level. The extraction of velocity values from experimental setup is done using PIVlab via the cross-relation conducted on the image data. A total of four different locations will be picked for the entire model to investigate changes in velocity along the steps. This point will indicate the performance of aeration efficiency with the existing design. Future improvements can then be applied based on the computed values from these points. The approximate locations of points of interest are picked within the steps and are illustrated by Figure 8 as below.

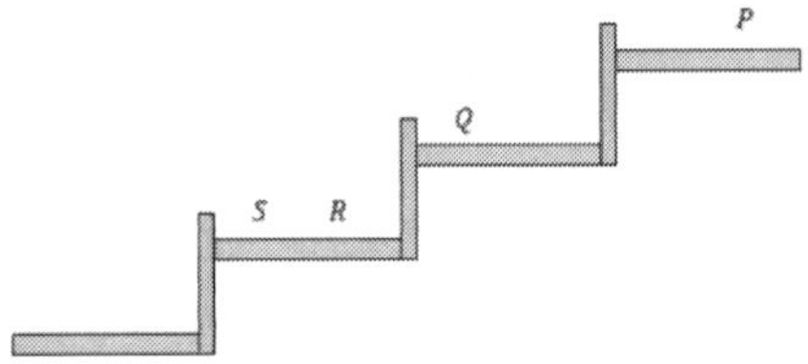

Figure 8: Points of interest located at each step for model

The points of interest and the corresponding velocities can be obtained from PIV image once image calibration is completed. Figure 9 summarizes the data obtained from experimental.

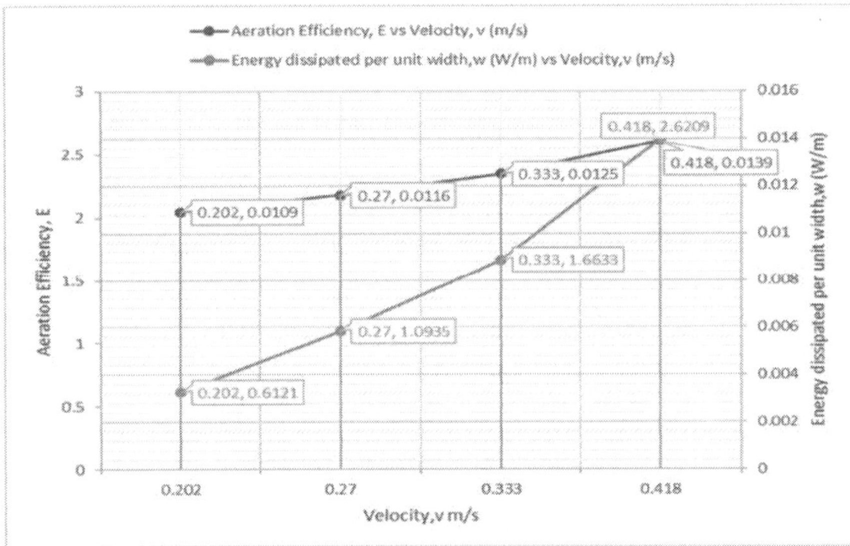

Figure 9: Plotted Graph from data collected

According to figure 9, the aeration efficiency of 0.0139 and a velocity value of 0.418 m/s was computed at point S which is highest compared to point R with aeration efficiency of 0.0125 and a velocity value of 0.333 m/s. The results show that the aeration efficiency increases as the velocity increases. These results occur because the velocity increase as the step size increases due to the potential energy from the top of cascade aerator changes into kinetic energy when it goes to the lowest step size. The lowest efficiency occurs at point P with aeration efficiency of 0.0109 and a velocity of 0.202 m/s at the top of cascade aerator. This lowest efficiency can be improved by increasing the barrier height to generate higher turbulence flow and swirl effect. Besides that, by increase the size of step also can affect the aerator efficiency due to the velocity increase.

CONCLUSION

Physical study was performed on a river bank configuration comprising barrier height of 10mm. The focus of the study was on the flow patterns, velocity and aeration efficiency performance. The stepped configuration showed identical flow pattern with nappe flow regime developed, which can be viewed as the discharge of water in a succession of free falling nappes from one step pool to the following. The results also highlighted an increasing of velocity pattern from point P → Q → R → S with a value of 0.202 m/s, 0.27 m/s, 0.333 m/s, 0.418 m/s. In terms of the energy dissipation and the aeration efficiency, similar increasing pattern is recorded for points P, Q, R and S with values of 0.6121 W/m, 1.0935 W/m, 1.6633 W/m and 2.6209 W/m while for the aeration efficiency with values of 0.0109, 0.0116, 0.0125 and 0.0139 respectively. This increasing pattern for velocity, aeration efficiency and energy dissipation indicated the importance of having cascade aerator design that with more height or increased in barrier height since it will promote air circulation thereby increasing the formation of manganese and iron oxides. As a result, cleaner water will be obtained towards the end of the cascade aerator system.

ACKNOWLEDGEMENTS

The authors would like to acknowledge Universiti Sains Malaysia and Uniten R&D Sdn. Bhd. for providing the facilities and financial assistance under accounts (304/PMEKANIK/60313052 and U-841).

REFERENCES

[1] H. Khdhiri, O. Potier, and J.-P. Leclerc, "Aeration efficiency over stepped cascades: Better predictions from flow regimes," *Water Res.*, vol. 55, pp. 194–202, May 2014.

[2] Amador, G. Van der Graaf, M. Sanchez-Juny, J.Dolz, F. Sanchez-Tembleque, J.Puertas, "Characterization of the flow field in a stepped spillway by PIV".

[3] Y. Peltier, B. Dewals, P. Archambeau, M. Pirotton, S. Erpicum, "Pressure and velocity on an ogee spillway crest operating at high head ratio: experimental measurements and validation", 2nd International Workshop on Hydraulic Structures: Data Validation, (2015).

[4] S. Felder, H. Chanson, "Aeration, flow instabilities and residual energy on pooled stepped spillways of embankment dams", Journal of Irrigation and Drainage Engineering, ASCE, Vol.139, No. 10, pp.880-887.

[5] X. Xie, C. Le Men, N. Dietrich, P. Schmitz, and L. Fillaudeau, "Local hydrodynamic investigation by PIV and CFD within a Dynamic filtration unit under laminar flow," *Sep. Purif. Technol.*, vol. 198, pp. 38–51, 2018.

[6] Q.-Q. Xiong, Z. Chen, S.-W. Li, Y.-D. Wang, and J.-H. Xu, "Micro-PIV measurement and CFD simulation of flow field and swirling strength during droplet formation process in a coaxial microchannel," *Chem. Eng. Sci.*, vol. 185, pp. 157–167, Aug. 2018.

[7] P. Guenther, S. Felder, H. Chanson, "Flat and Pooled Stepped Spillways for Overflow Weirs and Embankments: Cavity Flow Process, Flow Aeration and Energy Dissipation", (2013).

[8] Amador, M. Sanchez- Juny, J. Dolz, "Developing Flow Region and Pressure Fluctuations on Steeply Sloping Stepped Spillways", Journal of Hydraulic Engineering, 135(12), (2009).

AUTHOR INDEX

EDP Sciences – Web of Conferences
17, Avenue du Hoggar
Parc d'Activité de Courtabouf
BP 112
F-91944 Les Ulis Cedex A
France

ISBN 978-1-5108-7309-4

CURRAN ASSOCIATES INC.
proceedings.com